MAMMALS of Kentucky

MAMMALS
of Kentucky

Roger W. Barbour
& Wayne H. Davis

THE UNIVERSITY PRESS OF KENTUCKY

ISBN: 0-8131-1314-8

Library of Congress Catalog Card Number: 74-7870

Copyright © 1974 by The University Press of Kentucky

A statewide cooperative scholarly publishing agency
serving Berea College, Centre College of Kentucky,
Eastern Kentucky University, Georgetown College,
Kentucky Historical Society, Kentucky State University,
Morehead State University, Murray State University,
Northern Kentucky State College, Transylvania University,
University of Kentucky, University of Louisville, and
Western Kentucky University.

Editorial and Sales Offices: Lexington, Kentucky 40506

To

Wilfred A. Welter, in memoriam
W. Gene Frum, in appreciation

Contents

Appendixes

Color Plates follow page 132.

Preface

This is the fifth volume in a series of introductory guides to the flora and fauna of Kentucky. Because of its central location and the great variety of habitats that exist within its borders, Kentucky has been blessed with a rich natural heritage. Unfortunately, as in many other states, this heritage is being diminished and seriously threatened by the pressures of population growth and industrial technocracy. This series of books, it is hoped, will serve to acquaint more people with that heritage, and to encourage its preservation.

With that overriding purpose in mind we have prepared this volume on the mammals of Kentucky. Basically, we have pointed it toward a general audience—for readers who are not trained in mammalogy, but might be interested in knowing more about our state's mammals. We have sought to present accurate information without unduly encumbering the text with technical language and matters that concern only the professional. Throughout, we have pointed out, explicitly or by implication, the real paucity of information about the mammals of Kentucky; we hope thereby to stimulate the acquisition of additional knowledge.

With the hope of encouraging the increasing use of the metric system of measurement we have used this system exclusively in this volume. We personally feel that the total usage of the system is long overdue. In deference to those who may wish to convert metric dimensions to the more cumbersome English ones, we have included a table of equivalents.

In a volume such as this, many individuals and institutions become involved; to them, named and unnamed, we express our gratitude.

We are indebted to the Kentucky Research Foundation and the Research Fund Committee of the University of Kentucky for several grants over the years that made possible the accumulation of much information on the mammals of Kentucky.

Distribution data on mammals in Kentucky were accumulated from many sources; we are particularly indebted to Woodrow Barber, M. D. Hassell, Thane Robinson, Herbert Shadowen, and Pete Thompson who sent us information on the collections in their care at Morehead State University, Murray State University, the University of Louisville, Western Kentucky University, and Eastern Kentucky University, respectively. James Hardin sent us data from the collections of Southern Illinois University; Dave Fassler made his personal collection of Kentucky mammals available to us, as did Carl Ernst. Additionally, Dr. Ernst examined the Kentucky specimens in the U.S. National Museum for us. Joe Bruna and Jim Moynahan of the Kentucky Department of Fish and Wildlife Resources were especially helpful with information on the swamp rabbit and otter, respectively. Ray Nall and Paul Sturm of the Land Between the Lakes were most helpful with data on the fallow deer.

Mary DeLacey spent long hours in the library, tracking down distribution data, as did Faith Hershey. Additionally, Miss Hershey prepared the U.S. distribution maps, and typed much of the manuscript at least once. Julia Wade Smith helped us acquire photographs of the New England cottontail. John Whitaker supplied us with a color photograph of the southeastern shrew. Bernice L. Barbour did the track drawings and helped in the preparation of the Kentucky distribution maps.

Martha Jane Harrod, editor of *Kentucky Happy Hunting Ground,* allowed us to use separation negatives belonging to the Kentucky Department of Fish and Wildlife Resources for some of our color plates. The Tennessee Valley Authority supplied us with separations of a photograph of the fallow deer. The U.S. National Museum lent us mammal skulls to photograph.

All photographs not otherwise credited are by Roger W.

Barbour. Karl Maslowski and Marty Stouffer were especially generous of their beautiful photographs. Other photographers who contributed no less lovely but fewer numbers of photographs were Grady Franklin, Faith Hershey, John Hockersmith, Carl Kays, Alicia Linzey, John MacGregor, Bruce Poundstone, Delbert Rust, and Robert A. Stehn.

Our students, past and present, have stimulated us and kept us young—in outlook, if not in years.

To all of these, our thanks. It was you who made this book possible.

Table of Equivalents

Metric   *English*

LENGTH

1 centimeter(cm) = 10 millimeters(mm)	0.39(2/5) inch
1 decimeter(dm) = 10 centimeters	3.93 inches
1 meter(m) = 10 decimeters	39.37 inches
1 decameter(dcm) = 10 meters	10.93 yds.
1 hectometer(hm) = 10 decameters	109.36 yds.
1 kilometer(km) = 10 hectometers	0.62 mile

AREA

1 sq. meter	1.196 sq. yds.
1 hectare	2.47 acres
1 sq. kilometer	0.386 sq. mile

CAPACITY

1 liter(l)	1.10 dry quarts

WEIGHT

1 gram(g)	0.035 oz.
1 kilogram(kg) = 1000 grams	2.2 lbs.

TEMPERATURE

Centigrade = 5/9 (F − 32)
Fahrenheit = 9/5 C + 32

Introduction

About 200 years ago the land that is now Kentucky was invaded from the east by a population of light-skinned, scant-haired, bipedal mammals that have, in the short time they have been here, brought about more, and greater, changes in the fauna and flora than had ever been wrought by any combination of circumstances in any comparable period since the area last emerged from the sea.

We have no precise listing of the mammals inhabiting this area when white men first came, but we know that some species, common then, are no longer here. The bison, elk, gray and red wolves, panther, and migratory red men are gone. Other species, such as the bear and the otter, are either extremely scarce or are gone. Some, such as the white-tailed deer and the beaver, were almost eliminated but have been restocked and are again common. The deer thrive on the kind of diversified agriculture practiced in much of Kentucky, and there are probably more white-tailed deer in Kentucky now than in Daniel Boone's time. By opening up the forests and fostering grasslands, man also opened the way for colonization by grassland mammals.

GEOGRAPHIC AFFINITIES

The present mammalian fauna of Kentucky is an interesting mixture of species, derived, as it were, from all points of the compass. Because it is centrally situated in the eastern United States, Kentucky has in large part the mammalian fauna typical of the region. There are, however, various elements from other areas.

Several widely distributed southern species of mammals— Rafinesque's big-eared bat (*Plecotus rafinesquii*), the eastern harvest mouse (*Reithrodontomys humulis*), the golden mouse (*Ochrotomys nuttalli*), the rice rat (*Oryzomys palustris*), the cotton rat (*Sigmodon hispidus*), and the swamp rabbit (*Sylvilagus aquaticus*)—reach the northern limits of their range in or barely north of Kentucky.

The distribution pattern of many species of animals and plants characteristic of the Deep South is along the coastal plain from about Virginia southward across the Carolinas and thence westward across Georgia, Florida, Alabama, Mississippi, and Louisiana into Texas, with a northward extension up the Mississippi River valley into eastern Missouri, southern Illinois (and sometimes southern Indiana), and western Kentucky. The cotton mouse (*Peromyscus gossypinus*) has just such a distribution.

The western species in our mammalian fauna comprise 2 distinct elements; but, except for 2 rodents, these western forms are neither abundant nor obvious. One western element, from the grassy prairies, is represented by the prairie vole (*Microtus ochrogaster*) and the prairie deer mouse (*Peromyscus maniculatus bairdii*). The other western element consists of a few species typically inhabiting rough, rocky, brushy areas. These are perhaps best represented by the Mexican free-tailed bat (*Tadarida brasiliensis*), the western big-eared bat (*Plecotus townsendii*), the spotted skunk (*Spilogale putorius*), and the coyote (*Canis latrans*).

The northern species in our mammalian fauna are also readily divisible into 2 elements. Kentucky lies at or near the southern limits of the ranges of a number of widely distributed northern species; this element is best represented by the woodchuck (*Marmota monax*), the meadow vole (*Microtus pennsylvanicus*), and the meadow jumping mouse (*Zapus hudsonius*), all of which are wide-ranging in Kentucky. The second element of northern species in Kentucky is representative of a rather widespread situation in the flora and fauna of the southern Appalachians. Many wide-ranging northern species normally reach the southern limits of their distribution well to the north of Kentucky, but often a narrow extension of the range runs southward along the mountain crests into the southern Appalachians. Being slightly to the west of the crest, we do not have as high mountains, and consequently we lack the full complement of such species. However, a few of them occur at the higher elevations of our southeastern mountains, especially on Big Black Mountain, our highest, in Harlan County. This element in our mammalian fauna is best represented by such animals as the masked shrew (*Sorex cinereus*), the cloudland deer mouse (*Peromyscus maniculatus nubiterrae*), Gapper's redbacked mouse (*Clethrionomys gapperi*), the woodland jumping mouse (*Napaeozapus insignis*), and the New England cottontail (*Sylvilagus transitionalis*). Some 10,000 to 15,000 years ago, during the last Ice Age, these species must surely have ranged widely across this general area. As the ice receded and the climate warmed, most places became too hot and dry for them to survive. However, a few remnants remain—especially along the higher mountains, where the climate is still suitable for their existence, but also in deep gorges and other sheltered spots where the climate remains cool and moist.

HISTORY OF MAMMALOGY IN KENTUCKY

1750–1894

Our knowledge of the mammals of Kentucky began to accumulate in the 1700s in the form of casual remarks in the letters and journals of early visitors and settlers. Highlights of this early period were the explorations and writings of Constantine Rafinesque, in the early 1800s, and his descriptions of the prairie mole and the twilight bat from Kentucky. John James Audubon worked in Kentucky in this period, and his mammal work here and elsewhere resulted in the publication of his and John Bachman's monumental 3-volume *Viviparous Quadrupeds of North America* in 1845–54. In 1874 F. W. Putnam discussed some of the mammals from Mammoth Cave. This early period culminated in 1894 with the publication of H. Garman's "Preliminary List of the Vertebrate Animals of Kentucky."

1894–1925

Richard Ellsworth Call, in 1897, essentially began this new era with his paper on the fauna of Mammoth Cave, and in 1910 A. H. Howell contributed data on some mammals of western Kentucky. However, most of the firm data obtained in this period is to be found in various numbers of the U.S. Department of Agriculture's "North American Fauna" series, authored by E. A. Goldman, A. H. Howell, H. H. T. Jackson, G. S. Miller, Jr., E. W. Nelson, and W. H. Osgood. The publication, in 1925, of W. D. Funkhouser's *Wildlife in Kentucky* supplied a logical end to this middle period.

1925–1974

The early part of this period again saw the appearance of Kentucky information in the "North American Fauna" series, under the authorship of A. H. Howell and H. H. T. Jackson. William J. Hamilton, Jr., published his "Notes on the Mammals of Breathitt County, Kentucky" in 1930. Vernon Bailey, with the assistance of Leonard Giovannoli, made a major contribution in 1933 with his "Cave Life of Kentucky," a work not limited to cave-dwelling species. In the early 1930s Wilfred A. Welter, a young Ph.D. from Cornell University, came to Kentucky to head the Department of Science and Mathematics in what is now Morehead State University. In the short span of his tenure there (he was killed in an automobile accident in 1939) he exerted more influence on the progress of mammalogy in Kentucky than any other person. This was accomplished in small part by means of his 1939 publication (with D. E. Sollberger), "Notes on the Mammals of Rowan and Adjacent Counties in Eastern Kentucky"; his greater contribution was the inspiration and training he imparted to his students who have gone on in the field. In 1938 J. K. Neel wrote a master's thesis, "Lower Howard's Creek: A Biological Survey," which had data on mammals. In the 1930s the U.S. National Museum sent a survey party, headed by Watson Perrygo, to Kentucky. This expedition resulted in, among other things, the discovery of the existence of the red-backed mouse in Kentucky (reported by R. Kellogg in 1939).

Little work on the mammals of Kentucky appeared in the 1940s. However, in 1941 appeared the first of many publications on Kentucky mammals by R. W. Barbour, culminating in the present volume. In 1942 B. P. Bole and P. N. Moulthrop, in a paper on the mammals in the Cleveland Museum of Natural History collection, gave some locality data from Kentucky.

A major contribution in the 1950s was Barbour's 1951 paper, "The Mammals of Big Black Mountain, Harlan County, Kentucky." In 1954 Larry Gale and R. A. Pierce first reported the coyote in Kentucky, and W. Goodpaster and D. F. Hoffmeister published on the life history of the golden mouse in Kentucky. In 1956 Barbour discussed the status of the southern bog lemming (*Synaptomys cooperi*) in Kentucky and described a new subspecies from central Kentucky.

Six master's theses at the University of Kentucky in the 1950s dealt with Kentucky mammals. W. L. Gault and T. C. Tichenor treated the mammals of the Inner Bluegrass region and of Jefferson County, respectively; D. Y. Campbell, A. A. Dusing, H. E. Shadowen, and C. E. T. Smith worked on *Peromyscus*, *Sciurus*, *Microtus*, and *Pitymys* (=*Microtus pinetorum*), respectively.

Although one faunal study (on the mammals of Breathitt County) was done in the 1960s, this decade reflected a change from survey-type studies to those on animal behavior. The availability of radioactive tags and portable survey equipment stimulated this type of study and also made it more productive. In 1962 H. W. Ambrose III studied *Microtus* movements in this manner, and several other people followed with work on other mammals.

The addition of Wayne H. Davis to the teaching staff at the University of Kentucky, in the early 1960s, generated much interest in the study of bats, culminating in 1969 with the publication of Barbour and Davis's *Bats of America*. The year 1967 was a banner year for the study of mammalogy in Kentucky. Four theses at the University of Kentucky dealt with

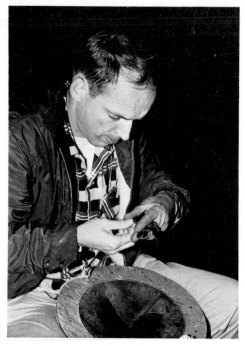

Roger W. Barbour (above) preparing mammal specimens at the summit of Big Black Mountain, Harlan County, July 1946, and Wayne H. Davis banding bats in Bat Cave, Carter Caves State Park, November 1963.

mammals: J. W. Hardin and M. D. Hassell both worked on the habits of bats in winter; M. J. Harvey completed a 3-year study on the prairie mole; and C. L. Rippy finished his study on the short-tailed shrew in Kentucky. In 1969 J. T. Wallace published on the golden mouse.

In 1970 Davis and Barbour published their findings on homing in blinded bats and J. K. Wade finished her study on the short-tailed shrew. In 1971 Wallace published his study on the meadow jumping mouse in Kentucky. D. J. Fassler has done extensive collecting in the vicinity of Somerset, and his paper on the mammals of Pulaski County is in press.

The appearance of the present volume, in 1974, brings to a logical end the third period in the study of Kentucky mammals. Many aspects of Kentucky mammalogy are still poorly known. We have endeavored in this volume to point out some of these deficiencies, in the hope of stimulating the necessary research to help fill the more glaring gaps.

PHYSIOGRAPHIC PROVINCES OF KENTUCKY

The accompanying map shows the physiographic provinces of Kentucky; the names shown are those used in this volume. For a generalized treatment of the physiographic provinces, the reader is referred to volume 3 of this series, *Kentucky Birds: A Finding Guide*, by Barbour et al.

THE SIGNIFICANCE OF WILD MAMMALS

There was a period in the early history of Kentucky when wild mammals supplied a major portion of the food of man— and a very minor part of his appreciation of the out-of-doors.

Now, fortunately, aesthetic appreciation of our wild mammals is widespread and growing, with corresponding decrease in the need for them as food. A magnificent deer, bounding gracefully along, is a joy many of us can savor again and again; when reduced to meat, the deer is enjoyed by only a few, and the little joy the food brings is soon gone forever. Even a tiny

PHYSIOGRAPHIC DIAGRAM OF KENTUCKY

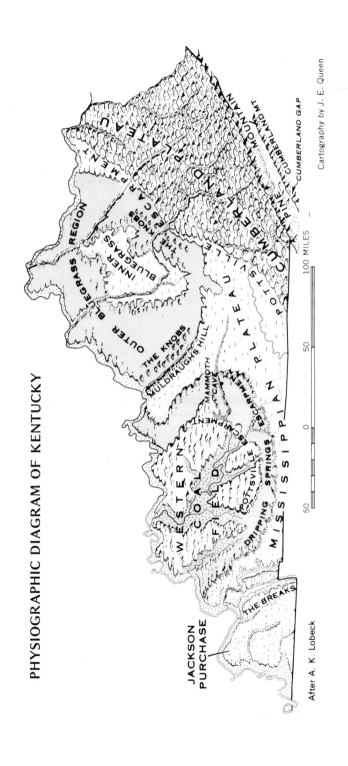

After A. K. Lobeck

Cartography by J. E. Queen

JACKSON PURCHASE

THE BREAKS

MISSISSIPPIAN PLATEAU

DRIPPING SPRINGS ESCARPMENT

POTTSVILLE ESCARPMENT

WESTERN COAL FIELD

MAMMOTH CAVE

MULDRAUGHS HILL

THE KNOBS

OUTER BLUEGRASS REGION

INNER BLUEGRASS

THE KNOBS

POTTSVILLE ESCARPMENT

CUMBERLAND PLATEAU

CUMBERLAND MT

PINE MOUNTAIN

CUMBERLAND GAP

100 MILES

50   0   50

shrew, scurrying about on the forest floor, is a wondrous sight, long remembered with pleasure.

A photograph, taken by one's own hand and skill, is a far better trophy than some dried and lifeless mounted head. Moreover, one animal can supply trophies for many people. There is no way as yet to reduce the aesthetic values of our mammals to dollars and cents, as one can readily do with the practical values of game species and furbearers. We are convinced, however, that aesthetic values are by far the greater.

Mammals, like all other animals, have a reproductive potential that is far in excess of the capacity of the land to support them. This means that many of the young are doomed to premature death. We have no objection to harvesting this surplus; in fact, we consider it desirable when done with care and moderation. One of the biggest problems is to determine what percentage of the population at any one time and place is really harvestable surplus. Fortunately, workers in the field of game management have come a long way in recent years, and the problems, although not completely solved, now seem capable of resolution.

Nine species of game mammals are hunted in Kentucky: all 3 of our species of rabbits, the woodchuck, gray and fox squirrels, the raccoon, and both the native white-tailed deer and the introduced fallow deer. Significance of these is treated in the species accounts.

The opossum, beaver, muskrat, red and gray foxes, raccoon, weasel, mink, and both of our skunks are trapped for their fur; in addition, an occasional bobcat is taken. The economic values of these are treated in the species accounts. The total value of the fur sold in Kentucky in 1972–73 was $289,058. The number of trappers in Kentucky is steadily declining, although prices paid for pelts are rising.

MAMMAL STUDY

Much valuable information about mammals can be accumulated by anyone by simple, direct observation and careful

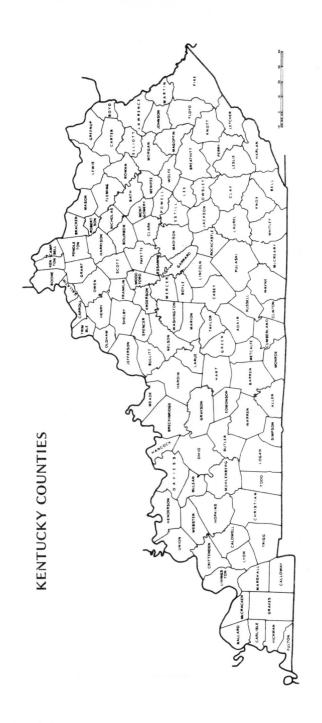

KENTUCKY COUNTIES

attention to note-keeping. For example, data on habitat, nesting habits, food, and territoriality can be derived from examining the arboreal nests and feeding platforms of golden mice. By carefully observing gray squirrels in a woodland (even a city park) one can establish many facts about intraspecific rivalry, breeding times, feeding habits, times of activity and their relation to climatic factors, and nests and nesting.

Many aspects of mammal study, however, require that the animal be captured alive and unhurt, for close examination or for marking in some manner so that the individual can be recognized at a later sighting. This requires live traps, which, for most purposes, are easily designed and constructed. On the other hand, some aspects of mammal study—taxonomy, anatomy, breeding cycles, food habits, and others—require that the animal be killed.

Small mammals can be readily taken in simple, snap mouse-traps, using a mixture of rolled oats and peanut butter as bait. The larger mammals can be taken by shooting or by means of various types of relatively humane traps. Also, much can be learned by examining road kills.

The smaller shrews are particularly difficult to take in standard traps, but a large tin can (No. 10) open at one end and sunk in the ground with the open top level with the surface is sometimes efficient.

Bats can be plucked by hand from a hibernating or maternity colony, or they can be taken in mist nets (the kind used by bird-banders). As a last resort, bats can be shot in the early evening in flight, using fine shot.

Keeping small mammals in captivity is reasonably easy, if attention is paid to details. Clean quarters, a constant supply of water, plenty of the right kind of food, and some sort of hiding place are essential.

Our wild mammals sometimes carry rabies, a serious disease that can be transmitted to man. Rabies most often occurs in the carnivores and bats; therefore people should especially try to avoid being bitten by these animals.

Frequently it is desirable to preserve a mammal for future

examination. This usually is accomplished by preparing a museum study skin and skeleton, but this requires considerable skill, which is attained only through practice. The simplest way to preserve a small mammal is to put it into Formalin (formaldehyde solution, usually 37–40%), which is available at drug stores and in school biology departments. Make a dilution of 10 parts of water to 1 part of Formalin; cut a longitudinal incision through the abdominal body wall of the animal, to facilitate entrance of the Formalin; and immerse the animal in the solution in an airtight container. If properly prepared, the animal will remain preserved indefinitely.

CONSERVATION

Many species of mammals in Kentucky are at least maintaining their numbers in most areas, but the creeping growth of strip-mining and other land-destroying practices, with the attendant destruction of animal and plant habitats, is posing serious problems for the native populations. Nearly 20% of our mammalian species appear on the list of rare and/or endangered species of wildlife in Kentucky, compiled by a committee of the Kentucky Academy of Science. Almost all of these are of limited distribution in the state and therefore could easily be eliminated from our fauna. If they are to survive, we must be careful not to destroy their habitats.

HOW ANIMALS ARE CLASSIFIED

By way of introduction to what follows, on the use of this book and the keys, the scheme of classification of animals is reviewed briefly here. The science of classifying organisms is taxonomy.

1. The animal kingdom is divided into major groups, called PHYLA (singular–phylum). There are about 40 phyla, but many of the commonly encountered animals—those possessing gill slits (in the embryonic stage or later) and a vertebral column or a notochord—are members of the phylum Chordata. This

Rare and/or Endangered Mammals in Kentucky
(For further information, see species accounts)

COMMON NAME	SCIENTIFIC NAME	STATUS
Masked shrew	*Sorex cinereus*	Rare
Southeastern shrew	*Sorex longirostris*	Rare
Southeastern myotis	*Myotis austroriparius*	Rare
Gray myotis	*Myotis grisescens*	Rare
Small-footed myotis	*Myotis leibii*	Rare
Indiana myotis	*Myotis sodalis*	Endangered
Townsend's big-eared bat	*Plecotus townsendii*	Rare
Cloudland deer mouse	*Peromyscus maniculatus nubiterrae*	Rare
Hispid cotton rat	*Sigmodon hispidus*	Rare
Kentucky red-backed mouse	*Clethrionomys gapperi maurus*	Rare
Woodland jumping mouse	*Napaeozapus insignis*	Rare
Eastern spotted skunk	*Spilogale putorius*	Rare
Bobcat	*Lynx rufus*	Rare & endangered
Coyote	*Canis latrans*	Rare & endangered
New England cottontail	*Sylvilagus transitionalis*	Rare

phylum includes all the fishes, amphibians, reptiles, birds, and mammals, as well as a few animals of more primitive kinds.

2. Phyla are divided into CLASSES. The bony fishes make up the class Teleostomi; the amphibians, the class Amphibia; the reptiles, the class Reptilia; the birds, the class Aves; and the mammals, the class Mammalia (discussed on page 17).

3. Classes are divided into ORDERS. The class Mammalia has 19 living orders, beginning with the order Monotremata, the egg-laying mammals; the order Marsupialia, the pouched mammals; the order Insectivora, the most primitive of placental mammals; and so on to the order Artiodactyla, the even-toed ungulates (pigs, deer, sheep, etc.). Some authorities arrange the orders differently, but there is general agreement on their composition.

4. Orders are divided into FAMILIES. For instance, the order Insectivora in Kentucky is composed of the families Talpidae, the moles, and Soricidae, the shrews. Note that the names of families end in -idae.

5. Families are divided into GENERA (singular—genus). For example, the family Soricidae is represented in Kentucky by one or more members of 3 genera: *Sorex*, the long-tailed shrews; *Blarina*, the short-tailed shrews; and *Cryptotis*, the least shrews.

6. Genera are divided into SPECIES (the word is both singular and plural). There are 3 species of *Sorex* in Kentucky: *Sorex cinereus*, the masked shrew; *Sorex fumeus*, the smoky shrew; and *Sorex longirostris*, the southeastern shrew. The species concept is difficult to define. Essentially, a species is a distinct kind of organism, in the way a mink is not a weasel although both belong to the same family and genus. Members of a species readily breed with each other but not, normally, with members of other species; if they do crossbreed, the offspring, called hybrids, are almost always sterile.

The scientific name of an animal species consists correctly of 3 parts; for example, *Sorex cinereus* Kerr. The first part (always capitalized) is the name of the genus. The second part (never capitalized) is the trivial name, or specific epithet. The third part (often omitted) is the name of the person who first described the animal and chose its trivial name. Often the describer's name is in parentheses; this simply indicates that, when he initially named the species, the generic name was not the same as it is now.

Note that the scientific name of a genus or a species is always printed in distinctive type, usually italic. It should also be mentioned that a species is sometimes referred to by genus only (for example, *Blarina*, standing for *Blarina brevicauda*) if only one species is being considered; and that the genus name may be abbreviated (as, *B. brevicauda*) after first use.

Groups of animals are called taxa (singular—taxon), which can be divided into subgroups, such as subphyla, subclasses, etc. These refinements are of little interest to the layman or the field biologist, except at the species level. A SUBSPECIES (also called a race) is a population that is somewhat different in appearance, anatomy, or behavior from other populations of the same species. A subspecies occupies a definite geographic area and intergrades with other subspecies where they meet. Obviously, 2 subspecies cannot long retain their identity within the zone where they interbreed freely; but if—rarely—one of the subspecies is prevented, for whatever reason, from interbreeding with the other over a long period, a new species may develop.

The name of a subspecies is formed by adding a term to the species name; thus, we have *Peromyscus leucopus leucopus* and *Peromyscus leucopus noveboracensis*, subspecies of the white-footed mouse. To save space and to give prominence to these names we may abbreviate, thus: *P. l. leucopus* and *P. l. noveboracensis* (provided the species name has first been given in full). Subspecies may also have vernacular names, such as prairie deer mouse and cloudland deer mouse (subspecies of the deer mouse).

HOW TO USE THIS BOOK

The arrangement of this book follows the taxonomic scheme outlined above, and the various taxa are organized according to a currently accepted scheme of evolutionary sequence; that is, the more primitive animals appear first, followed by those of increasing advancement.

The purpose of this volume is twofold: to enable a nonprofessional to identify any adult mammal occurring wild in Kentucky and to enable him to learn something of its distribution and life history. Identification may be accomplished in either of 2 ways:

1. By direct comparison with the color photographs, one can arrive at the proper identification in most cases and can come close to it in all. Then, by reading the species-account sections on recognition, variation, and confusing species, one can arrive at the correct identification. Partial verification can

be obtained by reading the sections on habitat and distribution and by checking the range maps.

2. Or one can use the illustrated keys. Verification of the identification can be obtained by comparing your specimen with the photographs and with the written accounts.

The keys are merely a series of numbered couplets, each offering a choice of characteristics headed *a* and *b*. By choosing the correct description, time after time, one is directed forward by number through the series of couplets until he arrives at the name of the animal in question. The small photographs with the key will be of value in making the correct decisions.

Following the discussion of the class Mammalia is a key to the orders of the class. When the proper order is chosen, one is directed to the page where the order is treated.

Following a description of the order is a key to the families represented in Kentucky. Here again, by making the proper choices, one is directed forward, to the treatment of the family. With a little experience and familiarity with the animals and the scheme of classification, one can turn directly to the proper family without bothering with the keys; instead, the Table of Contents can be consulted.

When the correct family is determined, the key presented there will enable one to identify the species (or, in a few cases, the subspecies). When the name of the animal is reached, one is directed to the page where the discussion of that animal occurs.

In the species accounts, and especially in the sections on distribution in Kentucky, frequent mention is made of the county names and physiographic provinces shown in the accompanying maps. The appropriate color plate is indicated for each species to the right of the common name.

Class Mammalia: Mammals

Mammals are the most highly developed of all animals and therefore are placed at the top of our system of classification. They are characterized by a number of features, most prominent of which are the presence of hair and of milk-secreting glands.

The term animal is frequently incorrectly used as a synonym for mammal, as in the expression "animals and birds." Birds are also animals, of course, and what is meant is "mammals and birds." Mammals are animals, but so are all other living things on this earth save plants and some organisms that belong to the shadowy worlds between plants and animals or between living and nonliving things.

Mammals are the ascendant animals on earth, if for no other reason than that we human beings are of this group. Most of our domesticated animals are mammals, and we are in no small measure dependent upon them for food and clothing.

In spite of their spectacular success, there is comparatively little diversity among the mammals. Worldwide, there are only about 1,000 genera and a little over 4,000 species, compared with, for example, more than 900,000 species of insects.

KEY TO THE ORDERS OF KENTUCKY MAMMALS

1. a. Innermost hind toe thumb-like, without a claw; tail prehensile, scant-haired; female with an abdominal pouch, in which the premature young develop; teeth 50: MARSUPIA-LIA, p. 19.

b. Inner toe of hind foot never thumblike, and with a claw; female without an abdominal pouch; teeth fewer than 50: 2

2. a. Forelimbs adapted for flight (bats): *CHIROPTERA*, p. 53

   b. Forelimbs not adapted for flight: 3

3. a. Toes armed with claws: 4

   b. Toes armed with hoofs: *ARTIODACTYLA*, p. 281

4. a. Front teeth (incisors) chisel-like and separated from the back teeth by a gap: 5

   b. Front teeth not essentially chisel-like and never separated from the back teeth by a gap: 6

4a

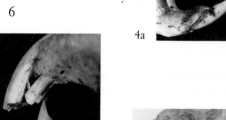

5a

6a

5. a. Tail short and fluffy; a pair of small, peglike incisors immediately behind the upper incisors: *LAGOMORPHA*, p. 120

   b. Tail various but never short and fluffy; upper incisors only 2, without a pair of peglike incisors behind the upper pair: *RODENTIA*, p. 133

6. a. Length generally less than 200 mm; eyes small, often im-
bedded in the fur; front teeth forcepslike: *INSECTIV-
ORA*, p. 26

   b. Length generally more than 200 mm; eyes large, well developed; front teeth never forcepslike: *CARNIVORA*,
p. 237

ORDER MARSUPIALIA   Marsupials

Marsupials constitute one of the most primitive of mammalian
orders, and they have a number of unique characteristics. The
young are born in an extremely underdeveloped state and require, at first, the protection of a pouch on the abdomen of the
mother.

These peculiar mammals are largely confined to Australia and
adjacent islands, but a few forms are found elsewhere. There
are several species in South and Central America, and one
species is distributed widely across the United States. The order
is represented in Kentucky by a single family, Didelphidae, and
a single species.

Virginia Opossum   PLATE 1

Didelphis virginiana (Linnaeus)

Recognition: Total length 650–835 mm; tail 275–350 mm;
hind foot 60–74 mm, weight 3–6 kg; males larger than females.
An animal about the size of a house cat, with a nearly naked,
scaly tail and naked ears. The inner toe on the hind foot is
thumblike and opposable. The female has a marsupial pouch
on the belly, in which small young are carried. Color usually

Distribution of *Didelphis virginiana* in the United States

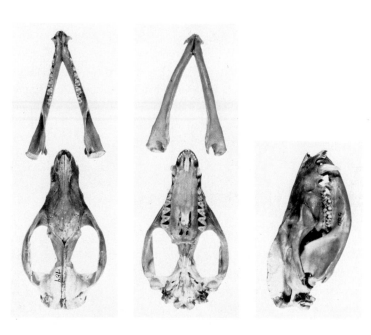

Skull of *Didelphis virginiana*, x 0.3

gray, but varies from black to reddish; the long guard hairs usually tipped with white.

Variation: No geographic variation in the opossum has been recorded in Kentucky. The subspecies here is *D. v. virginiana* Kerr.

Confusing Species: No other North American mammal can be easily confused with the opossum. The naked tail, naked ears, and opposable toe on the hind foot are a distinctive combination. The skull is easily recognized by the small braincase, the high sagittal crest, and the 50 teeth; no other Kentucky mammal has so many teeth.

Kentucky Distribution: Statewide. Abundant; less common at higher elevations.

Life History: The favored habitat of opossums is the forest edge and along woodland streams and ponds. They forage at night along creeks and gullies and, like other larger mammals, often use well-worn trails. By day they seek shelter in a wide variety of protected places, such as rock crevices, rock piles, woodpiles, cavities under buildings, hollow trees and logs, and burrows dug by other animals.

The opossum is one of the most omnivorous of mammals. Almost any kind of animal material and many fleshy fruits are eaten. Insects make up nearly half the diet, and carrion of all sorts is a major item. Fruit taken in season includes wild plums, crab apples, blackberries, wild cherries, and persimmons.

The opossum is nocturnal, normally arousing to activity as darkness approaches. It moves rather slowly and is easily captured by a person on foot. When cornered an opossum usually growls, opens the mouth wide in a threatening gesture, and erects the hair on its back. The animal can inflict a nasty bite when handled. Often, however, if one grasps an opossum and handles it roughly, it will feign death: the animal falls over, with its eyes and mouth partly open. If undisturbed it

Opossum playing "possum"

will often lie there for half an hour or so, after which it will raise its head and look cautiously about; if no danger is evident, it will get up and walk away.

When it is encountered when foraging at night, the opossum is relatively alert—not nearly so sluggish and stupid as it seems when encountered by day.

In the fall an opossum insulates its den with dry leaves, which it carries in bundles clutched in its tail. In gathering leaves, the opossum first takes them into its mouth and then, using its forepaws, passes them back beneath its body. It then puts its front feet on the ground, grasps the leaves with its hind feet, and places the leaves in a loop of the tail, in which they are carried to the nest.

Opossums do not hibernate, but in the coldest weather they sometimes remain denned up for several days at a time. They do not store food, and after several days of cold weather hungry opossums sometimes arouse and forage during the warmer daylight hours.

Opossums are usually solitary but show no indication of

intolerance or territoriality; the home ranges of individuals overlap. Occasionally 2 or more opossums may use the same den, but each goes its own way in foraging, and apparently the home ranges do not coincide.

When foraging an opossum travels an erratic route, seldom taking more than a few steps without a change of direction. Using its snout and paws, it pokes at objects on the ground, testing everything as a possible source of food.

Opossums are good climbers and often escape danger by taking to the trees. The prehensile tail is used in climbing; but, contrary to popular belief, an opossum does not normally hang by its tail, even though some opossums, especially the young, are able to do so. This can be demonstrated by putting a finger beneath the tail of a young opossum and lifting; the animal curls the tip of the tail about the finger and may be lifted from the ground.

The opossum is an accomplished swimmer and sometimes takes to the water to avoid capture. If frightened when swimming, an opossum may dive and then swim a short distance under water.

The young are born as early as January, but most early litters are produced in February or March. Some females produce a second litter.

Contrary to folklore, the female opossum is not impregnated through the nostrils; opossums breed like all other mammals. After a gestation period of only 11 to 13 days, a litter of a dozen or more honeybee-sized young is produced. As the young are born, the mother arches her body so that her head is within a few inches of her pouch, but she does not aid the young in their movements. As each baby emerges it immediately starts to crawl over the hair of the belly and into the pouch. Although 20 or more young are sometimes born, the mother has only 11 to 17 teats; those babies that do not find a nipple soon perish. As soon as a youngster takes a nipple in its mouth the nipple swells, filling the mouth so tightly that the baby becomes firmly attached to the mother. It remains so attached until it is 50 to 70 days old.

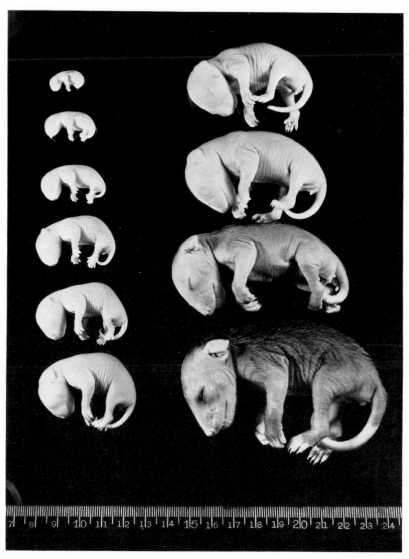

Young opossums. The smallest is less than 4 days old; the remainder
show growth at intervals of one week.

Mother opossum with 12-week-old young

As the young develop they become so large that they cannot all fit into the mother's pouch at one time; some then clutch the mother's fur and ride on her back as she forages. At about 90 to 100 days of age the young become independent and go their solitary ways.

Remarks: Although opossums become gentle in captivity and are easily kept, they do not make especially interesting pets. A baby that one of us rescued from the pouch of an opossum that had become a highway casualty was kept as a household pet for half a year. The animal seemed interested only in eating and sleeping. It would down food as rapidly as possible, using both front feet to stuff it in. A morsel such as a chicken leg, too big to be stuffed into the mouth, would be carried behind the sofa. Much of its active time was spent catching snails from the aquaria. The snails were captured with a front paw and placed in the mouth. As the opossum chewed, bits of snail shell would fall out of the corners of its mouth. Occasionally the opossum would lean too far over the tanks, tumble in, and have to be rescued. In its habits the opossum was remarkably clean. Droppings were deposited on a newspaper behind the sofa, and little evidence of its presence could be seen.

Through the southern states the opossum is seen dead on the highway probably more often than any other wild animal. It is so often a traffic victim because it is rather slow, seems to have poor eyesight, and feeds extensively on animals found dead on the highway. It seems remarkable that this primitive animal, which is so vulnerable to destruction by automobiles, dogs, and man, not only survives as an abundant animal but continues to extend its range. Opossums even invade our cities; it is not unusual to see one in downtown Lexington.

Opossums are eaten by man. If prepared properly they are delicious—especially when roasted with sweet potatoes.

Although opossum fur is of rather poor quality, the abundance of the animal makes its fur a rather important part of the annual harvest. In the 5 trapping seasons 1968–69 through 1972–73 fur-buyers in Kentucky purchased 27,829 Virginia opossum pelts—an average of 5,566 a year, but the number varied from 3,341 in 1970–71 to 9,355 in 1972–73. Average price per pelt was 59¢; the price varied from 36¢ in 1970–71 to 85¢ in 1972–73.

ORDER INSECTIVORA *Insectivores*

All our insectivores have a long, pointed snout, tiny eyes, and a skull that either lacks a zygomatic arch or has the arch much reduced. The teeth are sharp and pointed, and the canines are little differentiated from the premolars. The incisors project forward, and those of the upper and lower jaws, working together, form an excellent pair of forceps, admirably suited for picking üp insects and the like. Prominent scent glands are present in most species.

This order is widely distributed over the earth; the only extensive land areas that lack representatives are Australia and southern South America. It is represented in Kentucky by 5 genera in 2 families.

1. a. Forefeet enlarged and paddlelike, more than twice as broad as the hind feet; no apparent neck; zygomatic arch present: *TALPIDAE*, moles, p. 44
   b. Forefeet not broader than hind feet and not paddlelike; with an apparent neck; zygomatic arch absent; body mouselike: *SORICIDAE*, shrews, p. 27

Family Soricidae   Shrews

There are numerous structural and habitat differences between this family and the Talpidae, but one of the most striking is the small difference in size and structure between the forefeet and hind feet of the Soricidae. In the Talpidae the forefeet are much larger than the hind feet.

The family has essentially a tropical and temperate distribution; it is found throughout Europe, Asia, Africa, and North America, and in extreme northern South America. It is represented in Kentucky by 5 species in 3 genera.

KEY TO GENERA AND SPECIES OF KENTUCKY SORICIDAE

1. a. Tail short: less than 30 mm and generally about ¼ the length of head and body combined; ears hidden in fur: 2
   b. Tail longer: more than 30 mm and at least ½ the length of the head and body combined; ears visible (*Sorex*): 3

2. a. Total length usually less than 90 mm; only 3 unicuspids visible from the side: *Cryptotis parva*, Least Shrew, p. 41

b. Total length usually more than 90 mm; 4 unicuspids visible from the side: *Blarina brevicauda*, Short-tailed Shrew, p. 37

3. a. Total length usually more than 107 mm; pelage dark brown to gray: *Sorex fumeus*, Smoky Shrew, p. 35

   b. Total length less than 107 mm; pelage chestnut brown: 4

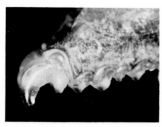

4a   4b

4. a. Total length usually more than 90 mm; third upper unicuspid not smaller than the fourth: *Sorex cinereus*, Masked Shrew, p. 28

   b. Total length usually less than 90 mm; third upper unicuspid smaller than the fourth: *Sorex longirostris*, Southeastern Shrew, p. 32

Masked Shrew; Common Shrew   PLATE 1

Sorex cinereus Kerr

Recognition: Total length 80–110 mm; tail 33–46 mm; hind foot 10–13 mm; weight 3.5–5.5 g. This is a tiny, slender, long-tailed, brown shrew with a long, sharp-pointed snout and minute eyes and with the ears nearly concealed in the fur.

Habitat of *Sorex cinereus* in Kentucky

Behind the upper incisors are 5 unicuspid teeth on each side; the fourth of these is generally smaller than the third, both are smaller than the first and second, and the fifth is minute.

Variation: No geographic variation is recognized in this animal in Kentucky; the species has been found at only one locality in the state. Our subspecies is *S. c. cinereus* Kerr. The color of the dorsal surface varies among individuals from pale brown to nearly black; winter pelage is darker than summer pelage. In younger animals the tail is somewhat furred; in old adults it is sometimes nearly naked. As the animal ages, the braincase flattens and broadens, the crests develop, and the upper incisors grow forward and downward.

Confusing Species: It is most similar to *S. longirostris*, with which it has often been confused, even by specialists. *S. longirostris*, a southern species, has a shorter tail (26–31 mm); a shorter, broader rostrum; and a shorter, more crowded unicuspid tooth row, with the third upper unicuspid smaller than the fourth. The darker *S. fumeus* is larger than *S. cinereus*. *Cryptotis parva* has a much shorter tail.

Kentucky Distribution: This is a northern species, known in Kentucky only from Big Black Mountain in Harlan County.

In Indiana and Ohio it has been taken in counties bordering the Ohio River, so it may also occur in northern Kentucky.

Life History: In Kentucky this shrew is known only from the deep, moist woodlands high on Big Black Mountain, where it prefers areas of thick leaf mold and decaying fallen trees. It is sometimes fairly common in such areas and, being active at any hour, can sometimes be seen by a careful observer.

This species is very prolific, with an average litter of 7, a gestation period of about 18 days, and, apparently, about 3 litters per season. In northern states pregnant females have been collected in every month from March to September. In spite of such fecundity, this shrew seems never to reach the abundance attained by meadow voles; and it usually is outnumbered by the short-tailed shrew, even in favored habitats.

The diet is varied but consists mostly of animal matter. Apparently these shrews will eat nearly any animal they can subdue. Insects, centipedes, snails and slugs, spiders, sowbugs, and various vertebrate remains have been found in their stomachs. Although vegetable matter makes up only about 1% of the food, these shrews may be taken in traps baited with oatmeal, peanut butter, or raisins.

Hawks, owls, cats and other large predators capture many shrews. The bones of this tiny species are often found in owl pellets. Some predatory mammals kill shrews but normally do not eat them.

Remarks: The common name, masked shrew, is unfortunate; we can find no indication of a mask.

Like other shrews, this species does well in captivity when properly cared for. Put lots of dirt and leaves into an aquarium and sink a water dish into the dirt. Feed the animal insects, meat, dog food, nut meats, rolled oats, etc.

On Big Black Mountain one of these intriguing little animals once took up residence, for a month or so, in a tent occupied by one of us, his wife, and their 2 small children. It lived

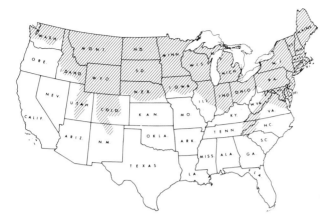

Distribution of *Sorex cinereus* in the contiguous United States

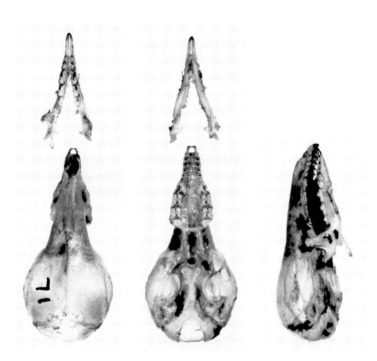

Skull of *Sorex cinereus*, x 3.2

under a storage box, and the family often saw it in early evening scurrying about on the tent floor.

This remarkable little shrew ranges to northernmost Alaska and is apparently active there throughout the winter. Metabolic demands in the Arctic winter must be enormous, and one wonders how such a tiny mite could find food enough to maintain itself. The breathing rate of this shrew has been measured at 850 per minute and the heart rate at 800 per minute.

Southeastern Shrew PLATE 2

Sorex longirostris Bachman

Recognition: Total length 72–108 mm; tail 26–31 mm; hind foot 10–13 mm; weight 2.5 g. This is a tiny, long-tailed, brown shrew.

Variation: No geographic variation is to be expected in Kentucky. Our subspecies is *S. l. longirostris* Bachman.

Southeastern shrew, *Sorex longirostris*

KARL MASLOWSKI

Distribution of *Sorex longirostris*

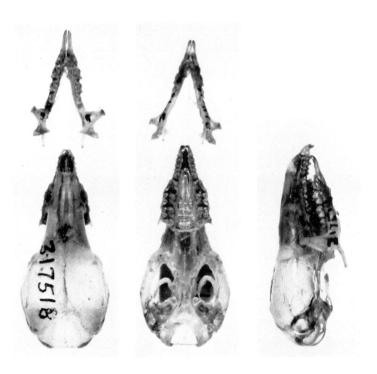

Skull of *Sorex longirostris*, x 3.6

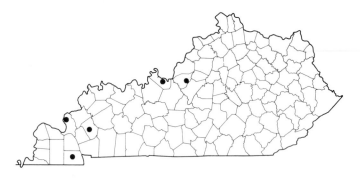

Locality records of *Sorex longirostris* in Kentucky

Confusing Species: This species is very similar to *S. cinereus*, from which it can be distinguished by the shape of the skull and the dentition. *S. cinereus* has a longer, narrower rostrum and a less crowded upper unicuspid tooth row. Also, the fourth unicuspid is smaller than the third; the reverse is true in *S. longirostris*. The larger *S. fumeus* is darker and has a longer tail. *Cryptotis parva* has a much shorter tail.

Kentucky Distribution: Western Kentucky and up the Ohio River Valley to Bernheim Forest in Bullitt County. Likely to occur in south central Kentucky but the limits of its distribution in Kentucky are unknown.

Life History: This is a rare shrew throughout its range. Its preferred habitat is swampy lowland weedfields, moist woods, and honeysuckle patches; however, a few specimens have been taken in rather dry upland fields. Little is known of its life history. The litter size is 4 or 5.

Remarks: This is one of the least known of all the mammals of eastern North America; either it is quite rare or our trapping methods are not effective for it. Probably the use of sunken cans, which has proven so successful in catching this shrew in recent years in Tennessee, would yield additional specimens in Kentucky.

Smoky Shrew

PLATE 2

Sorex fumeus Miller

Recognition: Total length 104–127 mm; tail 43–52 mm; foot 13–15 mm; weight 5–11 g. This is a rather large, dark, long-tailed shrew. Color varies from dark gray or nearly black in winter to gray tipped with brown in summer.

Variation: No geographic variation is evident in Kentucky. Our subspecies is *S. f. fumeus* Miller. Seasonal variation in pelage, from the darker cast in winter to some brownish tips to the fur in summer, is seen in all individuals. The tails of some individuals are entirely naked, but in most they are noticeably furred.

Confusing Species: The only other gray or black shrews in Kentucky are *Blarina brevicauda* and *Cryptotis parva*, both of which have short tails. All other shrews in Kentucky are brown, not gray, are pale tan below rather than gray, and are smaller than *S. fumeus*.

Kentucky Distribution: Scarce and irregular. Most often found on the Cumberland Plateau and in the southeastern mountains; however, specimens have been taken at the old entrance to Mammoth Cave, so the species may be locally

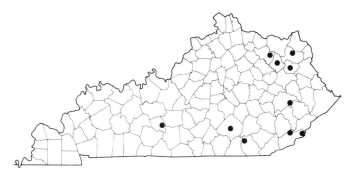

Locality records of *Sorex fumeus* in Kentucky

Distribution of *Sorex fumeus* in the United States

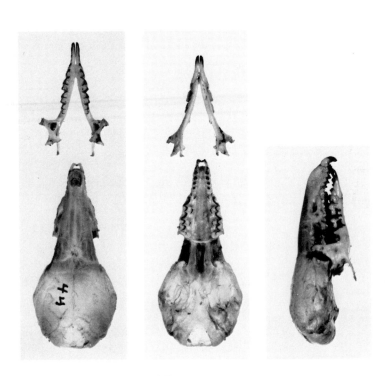

Skull of *Sorex fumeus*, x 2.7

distributed across southern Kentucky where suitable habitat is available.

Life History: The favored habitat of the smoky shrew is moist woodland where rocks, fallen trees, or brush provide adequate cover. We have also taken them along streams, in woodscrap at an abandoned sawmill, in weedy brush piles, and in swampy grassland. It seems that any place that provides adequate moisture, cover, and food may be inhabited by these shrews. Apparently because of its moisture requirements this species is local and never abundant in Kentucky; it is more common in the Appalachians, to the east.

Although smoky shrews will use the runways of other small mammals, they are most frequently caught in traps set among rocks or beneath a fallen log.

Apparently 3 litters are produced during the year; pregnant females have been taken from April to August. Litter size is 3 to 10, and the gestation period about 20 days.

Smoky shrews eat insects, earthworms, and whatever other small animal life they can find, as well as some vegetable matter. They are readily captured in traps baited with oatmeal, raisins, or nut meats.

Short-tailed Shrew PLATE 3

Blarina brevicauda (Say)

Recognition: Total length 95–134 mm; tail 17–30 mm; hind foot 10–17 mm; weight 15–30 g. A small, nearly black animal with a pointed nose, tiny eyes, short tail, and short legs and with the ears concealed in the fur.

Variation: Striking geographic variation is evident in Kentucky. In the northern and eastern regions a large race, *B. b. kirtlandi* Bole and Moulthrop, is found. Adults of this race usually measure 110 mm or more in total length. In the Purchase *B. b. carolinensis* (Bachman) occurs; this is a smaller

Distribution of *Blarina brevicauda* in the United States

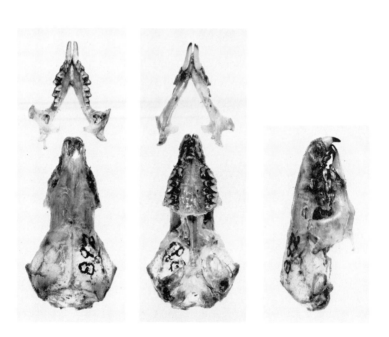

Skull of *Blarina brevicauda*, x 2.2

shrew, usually measuring 105 mm or less. The length of the hind foot is especially different, generally ranging from 15 to 17 mm in *B. b. kirtlandi* and from 10 to 12 mm in *B. b. carolinensis*. The two forms are so different that one suspects they may be separate species; however, they seem to intergrade in some parts of their range. In Nebraska *B. b. carolinensis* behaves as a distinct species. The situation in Kentucky is not clear at this time.

Confusing Species: Our only other short-tailed shrew is *Cryptotis parva*, which can be distinguished by its smaller size and its gray, rather than black, fur. *Blarina* is often mistaken for a young mole, and in fact it is commonly called "blue mole" in eastern Kentucky. Moles are easily distinguished from shrews, however, by their large, broad front feet.

Kentucky Distribution: Statewide; abundant. This is perhaps the most abundant mammal in Kentucky, being found wherever there is adequate vegetative cover, including city and suburban lots.

Life History: The preferred habitat of the short-tailed shrew is moist forest, but it also occurs commonly in brushland, brushy fencerows, weedfields, and even dense pastures. In any forest in Kentucky you can expect to find the 25 mm-wide runways beneath the leaf mold.

Like other small mammals *Blarina* varies markedly in numbers in different years. Populations as great as 625 per hectare have been recorded in Wisconsin. In late summer of 1972 a population "high" occurred in Kentucky. At that time short-tailed shrews were so abundant in suburban Lexington that they could often be heard squeaking or stirring among the leaves or beneath garden mulch. Once 3 were seen at one time on the lawn, seeking food and attempting to extend their runways from the protection of the hedges into the inadequate cover provided by the short grass.

In years of abundance shrews are sometimes heard rustling

among the leaves of the forest floor. At such times one can often detect the presence of *Blarina* by the characteristic odor of the secretions of the musk glands.

Like our other shrews, *Blarina* is active both day and night, summer and winter. Runways of moles and meadow voles, as well as their own, are used in quest of food to assuage their relentless appetite. They will take whatever is available, but they definitely prefer animal matter. Insects, earthworms, snails, slugs, centipedes, millipedes, spiders, and sow bugs are favored. Mice, salamanders, and the young of ground-nesting birds are sometimes taken. There are authentic records of these shrews attacking and subduing full-grown meadow voles in their natural environment. A salivary poison (which can give a person severe pain and swelling) helps the shrew overcome its prey.

Vegetable matter makes up about 20% of the shrews' food. This consists of fruits, roots, beechnuts, and acorns. They are attracted to oatmeal, peanut butter, raisins, and nut meats in traps. Food storage is apparently practiced, because collections of snails or paralyzed earthworms are sometimes found in shrew burrows.

Although generally terrestrial, short-tailed shrews occasionally climb in search of food. There is a record of one climbing 2 m up a small red oak to take suet from a bird feeder.

Feces of the short-tailed shrew are easily distinguished from those of mice. The pellets are much larger (nearly 25 mm long), spindle-shaped, and soft.

Breeding begins in February or March. Following a gestation period of 21 or 22 days, usually 5 to 7 young are born, in a bulky, crude nest of leaves, coarse grasses, and roots built in a burrow or beneath a log, stump, or rock. The pink, hairless young weigh about 1 g. They grow rapidly: they are furred at 10 days and about half grown at 2 weeks. They leave the nest at 18 to 20 days of age and are mature when 3 months old. Apparently the young do not breed until the following spring. The number of litters per season is unknown. Pregnant females that were still nursing have been captured.

Apparently because of its scent glands, *Blarina* is rejected by some predators. House cats often catch, but rarely eat, this shrew. Wild predatory mammals sometimes eat the hindquarters and leave the rest. Predators that do not disdain shrews include bass, trout, pike, rattlesnakes, copperheads, rat snakes, water snakes, great horned owls, and screech owls. The short-tailed shrew is probably the most beneficial small mammal in Kentucky. In addition to eating large quantities of insects and slugs, which plague the gardener, shrews may be in large part responsible for controlling the numbers of field mice. At a time of mouse abundance in New York, 56% of 200 *Blarina* scats examined contained mouse remains.

Least Shrew PLATE 3

Cryptotis parva (Say)

Recognition: Total length 70–89 mm; tail 13–18 mm; hind foot 9–11 mm; weight 4.2–5.6 g. A tiny, short-tailed shrew, brownish-gray to slate gray above and silvery below.

Variation: No geographic variation is recognized in Kentucky. Our subspecies is *C. p. parva* (Say). Winter pelage is somewhat darker than summer pelage.

Confusing Species: Our only other shrew with a short tail is *Blarina brevicauda*, which is larger, heavier, and darker and has 32 teeth, whereas *Cryptotis* has 30.

Kentucky Distribution: Statewide; fairly common. More common from the Bluegrass westward than in eastern Kentucky, where it becomes quite scarce and local.

Life History: The least shrew is an animal of the grasslands. It is most common in fields and fencerows that have a dense cover of ungrazed bluegrass.

Distribution of *Cryptotis parva* in the United States

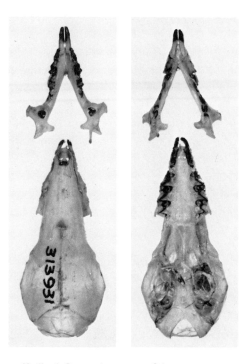

Skull of *Cryptotis parva*, x 3.1

These shrews are active day and night. They use the runways of meadow voles and pine mice, and sometimes they make their own tiny runs through the grass or beneath a board or log.

Seldom do least shrews fall to the efforts of the mammal-collector. They may be much more common than our collections indicate, however, for owl pellets sometimes contain the bones of many. Sunken cans are more effective for capturing them than are the more conventional traps.

Food consists of a wide variety of animals and plants. Beetles, bugs, crickets, grasshoppers, insect larvae, earthworms, and snails are favored. There are reports of this species' having nested in bee hives and feeding on the larvae. *Cryptotis* is also known to have attacked and eaten frogs. In captivity it will eat various kinds of fruits and vegetables.

The nest is spherical, about 100–125 mm in diameter, and composed of dry grasses or leaves. It is usually made in a depression in the ground or under a log, stone, or hay bale. Sometimes a nest is occupied by 6 or more adult shrews.

Perhaps the best way to find and capture these shrews is to locate a field in which scattered bales of hay have been left through the winter. Shrew nests are occasionally found under such bales, and the animals can easily be captured by hand.

Judging from data on captive animals, *Cryptotis parva* may be the most prolific of all mammals. Although the litter size is usually only 4 or 5, the gestation period is only 15 days, and another litter is on the way soon after one is born. Females have been known to give birth to, raise, and wean the litter and give birth to another litter—all within 24 days.

At birth the young weigh about 0.3 g and are blind, naked, and helpless. They grow very rapidly, attaining adult size by 3 weeks of age. A litter discovered beneath a board at Morehead in mid-April consisted of 6 young, which weighed a total of 17.6 g. The mother, still nursing the young, weighed 5.4 g, or less than a third of the combined weight of the young.

In the southern part of its range *Cryptotis* breeds throughout the year, but in Kentucky it seems not to be reproductively active in the winter months.

This is the easiest of our shrews to rear in captivity. It is a beneficial animal, and more people should become familiar with such a delightful little mite.

Family Talpidae Moles

Members of this family are remarkable for their extreme specialization for a subterranean existence. Some of their more outstanding adaptations for such an existence are the tremendously enlarged front feet and shoulder girdle, admirably fitted for digging; the short, soft fur, which lies equally well either forward or backward and thus enables the animal to more easily back up in its burrow; the cylindrical body, with a long, tapering snout; and the much-reduced pelvic girdle and hind legs, enabling the animal to turn with ease in a narrow passage.

The family ranges widely in North America, Europe, and Asia. It is represented in Kentucky by 2 species in 2 genera.

KEY TO GENERA AND SPECIES OF KENTUCKY TALPIDAE

1. a. Tail nearly naked: *Scalopus aquaticus*, Eastern Mole, p. 48

   b. Tail well furred: *Parascalops breweri*, Hairy-tailed Mole, p. 45

1a   1b

Hairy-tailed mole,
Parascalops breweri

Hairy-tailed Mole   PLATE 4

Parascalops breweri (Bachman)

Recognition: Total length 139–172 mm; tail 23–39 mm; hind foot 18–20 mm; weight 20–60 g. A black animal with large front feet about 12 mm broad and with a short, hairy tail. Eyes and ears are hidden in the fur.

Variation: No geographic variation is recognized in this species.

Confusing Species: Most similar in general appearance to the short-tailed shrew, *Blarina brevicauda*, which is smaller and does not have the large, nearly circular front feet of a mole. The forefeet of the short-tailed shrew are tiny, measuring only about 3 mm across, and are smaller than the hind feet. Shrews have brown-tipped teeth, in contrast to the white teeth of moles.

The common mole, *Scalopus aquaticus*, is larger than *P. breweri* and is easily distinguished by its essentially hairless, light-colored tail and the silvery sheen of its fur.

Kentucky Distribution: Eastern Kentucky; essentially limited to the Cumberland Plateau and the southeastern mountains. This is the common mole in the forested eastern part of the state.

Distribution of *Parascalops breweri* in the United States

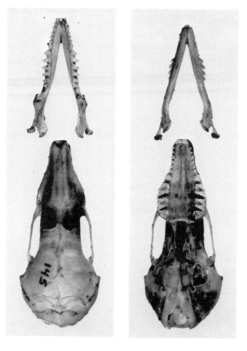

Skull of *Parascalops breweri*, x 1.5

Life History: This is a common animal of the eastern woodlands, where its characteristic raised tunnels can be seen crossing every forest path. Although it seems to prefer the woods, it is also common in pastures, lawns, and gardens. Where both species of moles occur together, the hairy-tailed mole tends to occupy the woods and the higher ground, whereas the common mole prefers open country and lowlands.

Although hairy-tailed moles prefer loose soils, they do not hesitate to enter hard-clay areas and will even tunnel under a dirt road. If the road is too hard-packed, however, the mole will sometimes end its tunnel on one side, cross the road on the surface, and re-enter the ground on the other side. In pastures, although they will forage into the grazed region, their main tunnels usually run along beneath the fence, where the ground is softer.

The hairy-tailed mole breeds in February and March. After a gestation period of about a month, 4 or 5 young are born, in a leafy nest about 150 mm in diameter in a burrow about 25–50 cm deep. The young are naked and helpless at birth but grow rapidly and are probably weaned within a month. Because moles (and shrews) apparently do not leave the nest until they are nearly full-grown, the young are not encountered unless one digs out the nest.

Moles feed avidly upon earthworms, insects, and other small animals. A captive hairy-tailed mole that weighed 50 g ate 66 g of earthworms and insect larvae in 24 hours. Moles do not hibernate; one wonders how they find enough to eat during the coldest weather, when invertebrates are scarce.

It is commonly believed that this mole eats potatoes, for often partially eaten potatoes are to be found where a mole run goes along the row. Actually, the damage usually is done by pine mice, which use the mole run as if it were their own. The pine mice are easily caught by digging into the mole run and setting a mousetrap, but a mole will almost invariably push such a trap out of the way or cover it with dirt.

Moles are active day and night and have no apparent pattern as to their times of activity; but they are rarely seen above ground. The extent of their activity in winter remains unknown. In the coldest weather they do not actively extend their surface feeding burrows, yet surely they have more difficulty finding food than in summer. Moles are known to bite and paralyze earthworms and store them as food, but the extent of this behavior is not known. It seems unlikely that such food would keep long enough to provide a winter store.

Hairy-tailed moles are economically neutral or somewhat beneficial; although they are sometimes a nuisance in lawns and golf courses, they eat many grubs and other destructive insects.

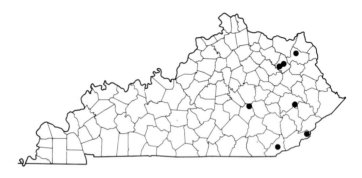

Locality records of *Parascalops breweri* in Kentucky

Eastern Mole; Bare-tailed Mole   PLATE 4

Scalopus aquaticus (Linnaeus)

Recognition: Total length 175–206 mm; tail 28–38 mm; hind foot 22–25 mm; weight 75–120 g. A robust animal with short legs; silvery-gray, velvety fur; large, broad front feet; and a short, nearly naked tail.

Variation: No geographic variation is recognized in Kentucky. The subspecies here is *S. a. machrinus* (Rafinesque), which was described from a specimen from Lexington.

Moles have pointed snouts and broad front feet.

Confusing Species: Easily distinguished from shrews by its larger size and broad front feet. The hairy-tailed mole, *Parascalops breweri,* is somewhat smaller and darker and has a hairy tail.

Kentucky Distribution: Nearly statewide; abundant, although local and less common in eastern Kentucky, where it is generally restricted to the valleys and is sometimes replaced by the hairy-tailed mole. Apparently absent from extreme southeastern Kentucky.

Life History: The common mole is an animal of loose, well-drained soils. Sandy soils of floodplains and streambanks, light loamy soils of the Bluegrass, meadows, pastures, gardens, lawns, and woodlands are occupied. Moles may persist in suburban housing tracts if the streets do not have curbs.

One of our former graduate students, Michael J. Harvey, studied common moles near Lexington. He found that they might be active at any hour of the day and night, summer and winter. However, activity peaks were from 4 to 7 a.m. and 6 to 9 p.m. In winter each mole used a single nest site, but in summer various nesting places were used.

Harvey found that the moles he studied were wide-ranging. A main run used by a mole may extend 275 m or more along a fencerow, with feeding tunnels branching off at favorable

Distribution of *Scalopus aquaticus* in the United States

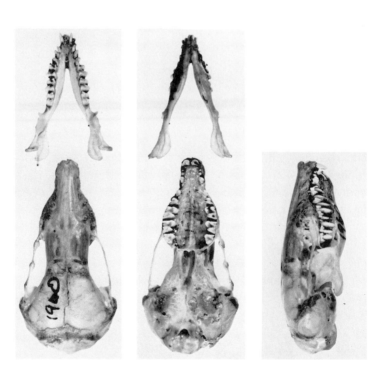

Skull of *Scalopus aquaticus*, x 1.4

Mole hills in light snow

localities. The home range varied from 1,512 to 3,430 m² for females and 3,616 to 18,041 m² for males. Sometimes the ranges of 2 individuals overlapped.

Two types of runs are used. The surface runs are easily located by the ridge of earth pushed up by the mole. Deeper runs—often 30 cm below the surface—usually serve as the main avenues of travel.

Common moles have a longer life and a lower reproductive rate than most small mammals. Although few mice live a year or more in the wild, Harvey recovered several adult moles 2 years after he had tagged them, and one was alive when he terminated his study, 3 years after its first capture.

After a gestation period of 45 days a litter of 2 to 5 is born, in April. Apparently a single litter is produced annually. Young moles are born in a bulky nest made of coarse grass interspersed with leaves and lined with finer grass. The nest is about 20 cm in length and 10 cm in diameter and is placed about 20 cm beneath the surface. At birth the young are rather large (about 50 mm long) and hairless. They grow rather slowly: they are more than a week old before hair appears and are only about half-grown at 5 weeks of age. Behavior of the young and the age at weaning are unknown.

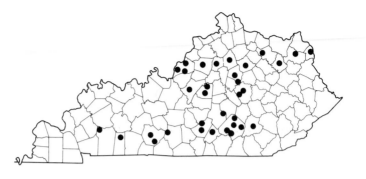

Locality records of *Scalopus aquaticus* in Kentucky

Food consists chiefly of invertebrates: earthworms, grubs, ants and ant pupae, slugs, snails, sowbugs, June beetles, spiders, centipedes, millipedes, and the eggs of insects, snails, and earthworms. A small amount of plant material is also eaten. Moles are reported to eat corn, potatoes, grass, tomatoes, apples, and cantaloupe. In captivity they will eat nut meats and rolled oats. Although moles have been reported to follow the rows of planted corn and eat the kernels and to go from one potato hill to another feeding upon the tubers, the real culprit in such instances in Kentucky usually is the pine mouse, which frequently uses mole runs.

The common mole very rarely comes above ground; it is seen on the surface even less often than the hairy-tailed mole. Occasionally it is driven to the surface by flood waters or emerges to cross a highway. There is a Michigan record of one seen traveling a surface run of a meadow mouse.

Moles seem to be most active after a warm rain. At this time they extend their surface runways, and the careful observer can sometimes catch them at work. They are very sensitive to disturbance as a person walks across the ground and will quit working and flee if one is not careful when sneaking up on the working mole. The mole can be caught by tramping down the burrow behind the place where the animal is working and then quickly digging it out.

Moles are easily captured in traps designed for taking them.

If one tramps down the dirt above a tunnel frequently used by a mole, the animal will push the dirt up again the next time it passes. Mole traps are designed to be triggered by this action. The mole cannot be captured by placing a trap over just any tunnel. Many tunnels made by a feeding mole are never used again. A good place to catch a mole is where his tunnel crosses a path or road or where it runs along beneath a fence. Only when moles become too disruptive of lawns, gardens, and golf courses should they be considered a nuisance. They are generally beneficial animals, feeding largely upon destructive insects.

ORDER CHIROPTERA Bats

Bats are the only mammals that possess forelimbs highly specialized for flight. The phalanges (finger bones) are much elongated and are connected by a leathery membrane, forming an efficient wing. Other outstanding characteristics of the order are the backward direction of the knee and the presence of a cartilaginous process, the calcar, arising from the ankle joint and providing support for the interfemoral membrane.

The ears generally are large and extremely acute. Our flying bats detect obstacles in their path by echolocation: they emit supersonic notes, hear the reflected sound, and thus are able to accurately place the reflecting object. Oilbirds, cave swiftlets, shrews, whales, porpoises, man (by his invention of sonar) and perhaps other animals also possess such a system of locating unseen objects. The eyes of our bats, although generally small, are used in navigation, in conjunction with echolocation.

Bats occur in all tropical and temperate parts of the world. They are represented in Kentucky by 8 genera in 2 families.

1. a. About half the tail projecting beyond the interfemoral membrane: MOLOSSIDAE, freetail bats, p. 117

   b. Tail inclosed in the interfemoral membrane, with no more than a few millimeters, at most, projecting free: VESPERTILIONIDAE, vespertilionid bats, p. 54

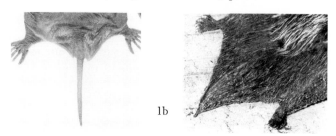

1a   1b

FAMILY VESPERTILIONIDAE   Vespertilionid Bats

Members of this family have simple (unadorned) muzzles. They all have complete interfemoral membranes, and in all of them the tail reaches to the back edge of the membrane but not more than a few millimeters beyond it.

Most of the small bats of northern latitudes belong to this family. In North America vespertilionids are found from the northern limit of tree growth south to Panama and the West Indies. Fourteen species in 7 genera occur in Kentucky.

KEY TO GENERA AND SPECIES OF KENTUCKY VESPERTILIONIDAE

1. a. Ears enormous: more than 26 mm from notch to tip: 2

   b. Ears less than 26 mm from notch to tip: 3

2. a. Belly fur white; first upper incisor bifid: *Plecotus rafinesquii*, Rafinesque's Big-eared Bat, p. 113

   b. Belly fur tan to brown; first upper incisor unicuspid: *Plecotus townsendii*, Townsend's Big-eared Bat, p. 109

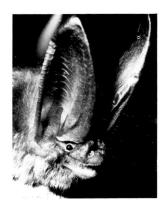

1a

1b

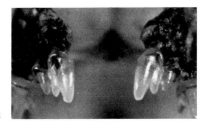

2a

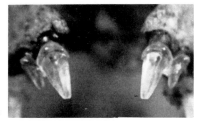

2b

3. a. At least the anterior half of the dorsal surface of the interfemoral membrane nearly as heavily furred as the rump: 4

   b. Dorsal surface of the interfemoral membrane naked, scant-haired, or, at most, lightly furred on the anterior third: 6

4. a. Color of fur black, but many of the hairs distinctly silver-tipped: *Lasionycteris noctivagans*, Silver-haired Bat, p. 85

   b. Color various but never uniformly black; hairs may or may not be silver-tipped: 5

5. a. Ears conspicuously black-edged and with patches of yellowish hair scattered inside them; color dark, hairs silver-tipped: *Lasiurus cinereus*, Hoary Bat, p. 103

   b. Ears not conspicuously black-edged, and bare or, at most, scant-haired on the inside; color red-orange or yellowish-brown: *Lasiurus borealis*, Red Bat, p. 99

6. a. From a side view, the first visible tooth behind the upper canine approximately ½ as high as the canine and in contact with it at the base: 7

   b. From a side view, the first visible tooth behind the upper canine less than ⅓ as high as the canine or, if ½ as high, then separated from the canine by a noticeable gap: 8

7. a. Forearm more than 40 mm long: *Eptesicus fuscus*, Big Brown Bat, p. 93

   b. Forearm less than 40 mm long: *Nycticeius humeralis*, Evening Bat, p. 106

8. a. Hairs on the back dark at the base and the tip, lighter in the middle: *Pipistrellus subflavus*, Eastern Pipistrelle, p. 89

3a

5a   5b

6a

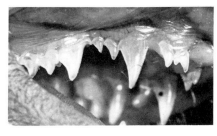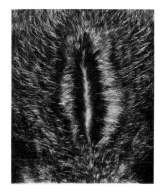

6b

8a

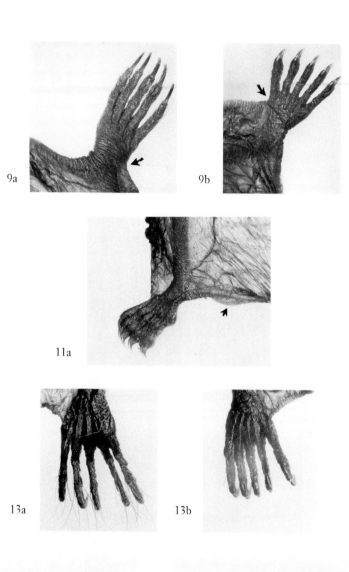

9a   9b

11a

13a   13b

14a   14b

b. Hairs on the back either dark at the base and lighter at the tip or else uniformly colored; no light band in the middle (genus *Myotis*): 9

9. a. Hairs on the back uniformly colored from base to tip; wing attached to the foot at the ankle: *Myotis grisescens*, Gray Myotis, p. 69

   b. Hairs on the back darker at the base than at the tip; wing attached to the foot at the base of the toe: 10

10. a. Ear when gently laid forward extending more than 2 mm beyond the tip of the nose: *Myotis keenii*, Keen's Myotis, p. 72

    b. Ear when gently laid forward extending less than 2 mm beyond the tip of the nose: 11

11. a. Calcar with a keel: 12

    b. Calcar not keeled: 13

12. a. Foot more than 8.5 mm long; forearm usually more than 35 mm long: *Myotis sodalis*, Indiana Myotis, p. 76

    b. Foot less than 8.5 mm long; forearm usually less than 35 mm long: *Myotis leibii*, Small-footed Myotis, p. 81

13. a. With a few scattered long hairs on foot extending to the tip of the claws or beyond: 14

    b. No long hairs on the foot extending to the tip of the claws: *Myotis sodalis*, Indiana Myotis, p. 76

14. a. Dorsal fur dull, not glossy; frontal area of the skull rising abruptly from the rostrum; sagittal crest usually present; braincase viewed from the side almost spherical: *Myotis austroriparius*, Southeastern Myotis, p. 65

    b. Dorsal fur usually with a conspicuous sheen; frontal area of the skull rising gently from the rostrum; sagittal crest lacking; braincase viewed from the side flattened, not spherical: *Myotis lucifugus*, Little Brown Myotis, p. 60

Little Brown Myotis;
Little Brown Bat PLATE 5

Myotis lucifugus (Le Conte)

Recognition: Total length 85–108 mm; tail 37–48 mm; foot 10–11 mm; ear 14–16 mm; wingspread 222–272 mm; weight 6–13 g. A medium-sized brown myotis, usually with sleek, glossy fur. Long hairs on the toes extend beyond the tips of the claws.

Variation: No geographic variation is recognized in Kentucky. The subspecies here is *M. l. lucifugus* (Le Conte). Color varies from pale tan through reddish-brown to dark brown. A few individuals in summer have short fur that is not glossy.

Confusing Species: Most similar to *M. sodalis,* from which it can be distinguished by the long hairs on the toes. Also confused with *M. keenii,* which has longer ears and a longer, more pointed tragus. *M. austroriparius* has dull, woolly fur and a higher, more nearly spherical braincase. *M. leibii* is smaller and has a much smaller foot. *M. grisescens* is larger, has

Little brown bat, *Myotis lucifugus,* caught in a mist net

Ear of *Myotis lucifugus* showing tragus. The tragus of *M. keenii* (p. 73) is longer and more pointed.

monochromatic fur, and has the wing membrane attached at the ankle. *Nycticeius humeralis* has a short, rounded tragus.

Kentucky Distribution: Statewide. Common in winter in many of our larger caves. In summer it is common but local in nursery colonies, tending to be more frequent in the eastern and northern parts of the state. A few wandering males can be taken in summer by netting at the mouth of nearly any major cave or mine. In the Purchase it is scarce, known from a single specimen from Murray.

Life History: The little brown myotis is a northern species, reaching its maximum abundance in the northern tier of states. Kentucky is near the southern limit of its summer range.

In summer this bat inhabits buildings, usually choosing a hot attic and there establishing a nursery colony of hundreds—even thousands. Less frequently, colonies are found beneath tar paper, siding, shingles, and similar sheltered spots.

Colonies are usually close to a lake or stream. If water is not close by, the bats forage among trees in rather open areas, such as along tree-lined streets.

Nursery colonies begin to form in April, soon after the hibernating groups begin to disperse. In Kentucky the major hibernating groups—in Carter Caves State Park and in the Mammoth Cave region—migrate northward to summer homes in Ohio and Indiana. Caves in Lee County, however, serve as wintering grounds for nursery colonies in nearby St. Helens and Clay City.

Bats from a large summer colony that occupies the post

Distribution of *Myotis lucifugus* in the contiguous United States

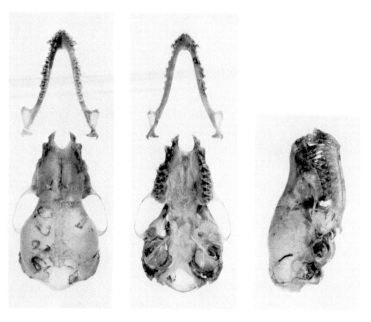

Skull of *Myotis lucifugus,* x 3

office and a nearby building at Oakville, in Logan County, have been found in winter in caves near Nashville, Tennessee. *M. lucifugus* normally breeds in the fall, but occasionally a mated pair is found in winter. In any large hibernating colony some bats are active at any time, and it is not unusual to find an active pair mating. After copulation, sperm are stored in the uterus through the winter; fertilization occurs when the bats emerge from hibernation. Gestation takes 50 to 60 days. Parturition occurs over a wide time-span: in Kentucky a few young are born as early as May 21 and an occasional female is still pregnant on June 21. Yearlings give birth later than older adults, and many yearling females do not bear at all. A single young is produced each year.

The baby is very large at birth, weighing 1.5 to 1.9 g—about one-fourth the weight of the postpartum mother. It is born blind, but its eyes open on the second day. The baby clings tightly to a nipple during its early days of life. When at rest by day, the mother keeps the baby beneath a wing. If disturbed she takes flight, carrying the baby attached to a nipple. As a result of disturbance she may move her baby to another colony in the neighborhood.

The young grow rapidly and begin to fly about in the roost when 3 weeks old. At age 1 month they have attained adult weight and begin to fly outdoors, where they learn to catch insects on the wing. They must learn to detect insects by echolocation, to distinguish them from such things as twigs and leaves, and to capture them in the wing and tail membranes. If the young are to survive they must become such skillful hunters that they can store fat for the long hibernation season. Because the young are less effective as food-catchers than are the adults, they carry less fat at the start of hibernation and consequently have a higher mortality in the first winter.

Throughout most of the species' range the sexes are segregated in summer; males lead a solitary life, and females congregate in nursery colonies. In Kentucky, however, we find many males in these colonies: 21% to 34% of the adults. Clustering in attics is apparently a means of maintaining high

Little brown bats, *Myotis lucifugus*, hibernating in a Kentucky cave

temperature and, consequently, rapid growth. In a colony at Oakville on August 1, 1963, the temperature was 55°C—the highest temperature ever known to be tolerated by any small mammal.

By August the bats have put on so much fat that it amounts to about ⅓ of their body weight. Soon the colonies break up, and the bats begin to wander toward their wintering caves. In August and September they appear by the hundreds, along with several other species of bats, at several of our major caves, where they swarm about, in and out of the caves. A few sometimes remain in certain caves by day, but the daytime whereabouts of most of these bats is unknown.

With the arrival of cold weather the bats go into hibernation. Suspended from the wall or ceiling of a cave, they sleep through the winter, except that they awaken every 2 weeks or so. At these wakeful times they may seek a more favorable spot or sample the weather outside.

M. lucifugus has a remarkable life-span, which is necessary to maintain its numbers, in view of the very low reproductive rate. Several individuals are known to have lived 20 years in the wild, and one was recaptured 24 years after being tagged.

Predators, accidents, and man are the main causes of death

in little brown bats. House cats often capture them in summer. Mink and raccoons enter caves and eat them. Rat snakes will climb onto the rafters of buildings to feed upon bats. One little brown myotis we had banded in Kentucky was recovered from the stomach of a bass taken from Nolin Reservior. Voles, white-footed mice, trout, and frogs have all been reported as eating little brown bats.

After a flood in a Kentucky cave, hundreds of these bats were found drowned. There are several reports of these bats becoming entangled in dried burs of the burdock and perishing there, and one was once found impaled on a barbed-wire fence.

Man is continually fighting bats that take up residence in houses. When a large colony is present they can be a genuine nuisance: their odor can permeate the entire house. The bat bug (*Cimex*), a close relative of the bedbug, becomes abundant in summer colonies, and when the bats leave in the fall some of these bugs wander downstairs—hence the tale that bats carry bedbugs.

Like all other species of bats in Kentucky, *M. lucifugus* feeds entirely upon flying insects; so, if they are not causing annoyance in one's home, they are good citizens to have around. Unfortunately, most species are rapidly declining, apparently as a result of man's activities. The causes are not precisely known, but disturbance of the hibernating colonies and poisoning with DDT and other pesticides are the most likely factors. Bats store pesticides in their fat in autumn and become very susceptible to the toxic effects the following spring.

Southeastern Myotis;
Southeastern Bat   PLATE 5

Myotis austroriparius (Rhodes)

Recognition: Total length 87–99 mm; tail 37–45 mm; foot 10–11 mm; ear 14–16 mm; wingspread 238–270 mm; weight 7–12 g. A medium-sized myotis with a high, rounded brain-

Distribution of *Myotis austroriparius*

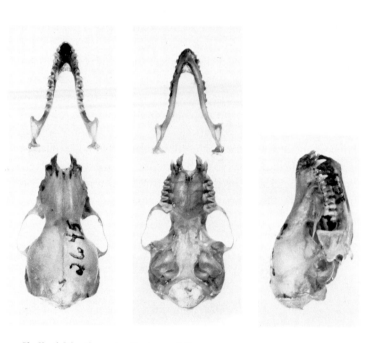

Skull of *Myotis austroriparius*, x 2.8

case and, usually, a slight sagittal crest. The fur has a woolly appearance, and there is little contrast between the base and the tip of hairs. Long hairs on the toes extend well beyond the tips of the claws. Color russet to gray above, tan to whitish below.

Variation: No geographic variation is recognized. New pelage in the fall and winter is gray above and white below. In summer many individuals become various shades of brown and russet. (This is a common pattern in many species of colonial bats; it is caused by the ammonia fumes from bat urine.) In western Tennessee many specimens taken while molting in August have patches of both gray and russet fur.

Confusing Species: Most similar to *M. lucifugus*, from which it can be easily distinguished only by examining the cleaned skulls or the nearly microscopic penis bones. The braincase of *M. lucifugus* is lower and less spherical. The fur of *M. austroriparius* lacks the glossy quality generally seen in *M. lucifugus*. For comparisons with other similar species see the account of *M. lucifugus*.

Kentucky Distribution: Known only from 2 caves in Mammoth Cave National Park, where occasional stragglers have been found in summer and winter, from a cave near Hopkinsville, and from a cave in Livingston County. The species is probably common in the Jackson Purchase, in western Kentucky, because it is quite common in the neighboring counties of western Tennessee.

Life History: This is a bat of the Deep South that comes up the Mississippi River valley and reaches the limits of its range in southern Illinois and Indiana and in western Kentucky. In these states it is known almost exclusively from specimens taken in caves, mostly in winter.

Throughout most of the species' range there are no caves; there the bats are generally associated with quiet water, where they forage just above the surface. In western Tennessee this species is commonly taken in nets placed over a tree-lined stream or a woodland pond.

Little is known about the habits of this bat in the northern part of its range. Nursery colonies should be sought in buildings and hollow trees throughout the lowlands of western Kentucky.

This is our only myotis that produces more than one young per litter: 90% of the pregnant females have twins; the remainder have a single young. In Florida, where the only studies on reproduction have been done, parturition begins in late April and peaks about the second week in May. Yearlings of both sexes are reproductively active, and breeding apparently occurs in spring.

In the Florida caves a few adult males are found in the large nursery colonies, but most are scattered in other shelters, singly or in small groups. After the young mature, many adult males join the colony. In October the colonies disperse.

The life history of M. *austroriparius* in Kentucky probably is somewhat different from that in Florida. When a nursery colony is finally discovered in Kentucky or a neighboring state, much interesting information about this little-known bat can be obtained.

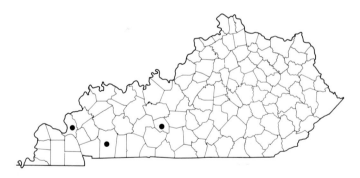

Locality records of *Myotis austroriparius* in Kentucky

Gray Myotis; Gray Bat   PLATE 6

Myotis grisescens (Howell)

Recognition: Total length 97–106 mm; tail 38–45 mm; foot 12 mm; ear 16 mm; wingspread 275–300 mm. A large, big-footed myotis, with the wing membrane attached to the foot at the ankle. Calcar not keeled; a distinct sagittal crest on the skull. Color uniformly gray from base to tip of hairs, but occasional individuals are russet in summer.

Variation: No geographic variation is recognized. Color varies slightly from light gray to dark gray. In summer there is some bleaching of the fur, so that some individuals are brown to russet.

Confusing Species: Similar to all our other *Myotis*, but larger. Easily distinguished, because it is the only one with the wing membrane attached to the ankle instead of the base of the toes and because it is our only bat in which the dorsal hairs are the same color from the base to the tip.

Kentucky Distribution: The cave region of south-central Kentucky. Within the past 10 years there have been nursery colonies in caves in Trigg, Adair, Taylor, Garrard, and Jessamine counties and a major winter colony in Edmonson County. A specimen was taken in Bat Cave, in Carter County, in 1931, and stains on the wall suggest that a large colony may formerly have inhabited this cave. This is one of our most rapidly vanishing wildlife species, and its status is changing from year to year. It occurs in Kentucky at this writing, but its future is in serious doubt. No colony in this state is protected.

Life History: This is a cave bat. There are a few records from man-made structures, such as tunnels and storm sewers, and a recent report of a nursery colony in a barn in Missouri; but in Kentucky all our records are from caves.

This species is especially intolerant of disturbance by man.

Distribution of *Myotis grisescens*

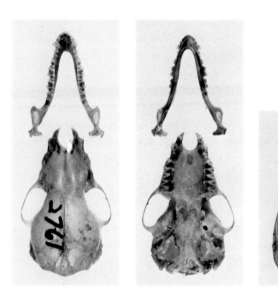

Skull of *Myotis grisescens*, x 2.6

Once abundant in hundreds of the caves of Tennessee, Arkansas, Alabama, Missouri, and Kentucky, it has now disappeared from most of those that are easily accessible to man. The caves still harboring gray bats in summer are mostly those accessible only by boat or by wading deep water. The major winter colonies of this bat—colonies unknown to science as recently as 20 years ago—are hidden away in caves that can be reached only via deep pits. Nearly the entire population winters in fewer than a dozen such caves; each harbors about 100,000 bats. One of these is in Kentucky: the Coach-James cave system, just outside Mammoth Cave National Park. It is one of several of these bat caves that have been recently commercialized, and the fate of the bats is unknown. Unfortunately, none of the several dozen caves in Mammoth Cave National Park harbors a colony of *M. grisescens*, although transient individuals can be found in a few of them.

The boom in the hobby of cave exploration, together with the increasing sophistication of equipment and techniques for entering the most challenging caves, poses a serious threat to the survival of the gray bat. Except for a state park in Florida and a cave-research laboratory in Missouri, we know of no colony receiving any protection. The U.S. Forest Service is attempting to protect a winter colony in a cave in Arkansas. Although the populations in some other states will probably outlast those in Kentucky, we think this species is headed for extinction.

In Kentucky the bats begin to assemble in the nursery caves in late March and early April. Adult males arrive with the females; sometimes they make up as much as 20% of the colony. In June each pregnant female gives birth to a single young. In Arkansas the young are mostly born in the first half of June. A study in Missouri found that yearling females did not produce young; they segregated with the males in a part of the cave away from the nursery.

After the young are weaned, the colonies begin to break up; in August many are deserted. The bats wander great distances at this time and can be captured by netting at night at various

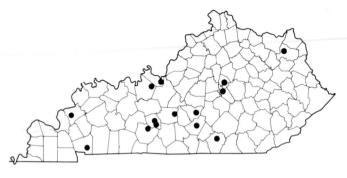

Locality records of *Myotis grisescens* in Kentucky

large caves. They attain their maximum amount of fat in October after arriving at the hibernating sites. Breeding occurs at the hibernation site in late fall. Sperm are retained in the uterus, and fertilization occurs after the bats emerge, in late March. The bats sometimes travel long distances to reach their summer caves.

Keen's Myotis; Keen's Bat PLATE 6

Myotis keenii (Merriam)

Recognition: Total length 86–98 mm; tail 37–45 mm; foot 10 mm; ear 17–19 mm; wingspread 228–258 mm; weight 5–10 g. A medium-sized myotis with long ears and a narrow, pointed tragus. Calcar not keeled; fur brown, not glossy.

Variation: There is limited color variation, ranging from medium brown to dark brown. The subspecies in Kentucky is *M. k. septentrionalis* (Trouessart).

Confusing Species: Very similar in general appearance to *M. lucifugus*. It can easily be distinguished from this and all of our other *Myotis* by its longer ears and especially the long, narrow, pointed tragus.

Ear of *Myotis keenii* showing the long pointed tragus. For comparison with *M. lucifugus* see page 61.

Kentucky Distribution: Found sparingly throughout the state except the Jackson Purchase, in western Kentucky. Scarce and local in and about certain caves.

Life History: We do not understand the distribution and abundance of this bat in Kentucky. It is a permanent resident but is nowhere common. In years of working with the hibernating bats in Carter Caves State Park and netting thousands of bats there in the warm months, we never found this species. Yet at nearby Tar Kiln Cave, in Elliott County, in netting for only 2 hours on the night of August 5, 1964, we captured 6 male *M. keenii,* and others of this species were seen in the cave.

Likewise, in Mammoth Cave National Park, where we netted and banded more than 12,000 bats in the late summer of 1963, we caught fewer than half a dozen *M. keenii.* However, James Cope and his students, netting at the same time at a cave in nearby Meade County, regularly captured this species in small numbers.

In Crittenden County, in western Kentucky, in June 1957 we collected 5 males of this species by shooting them as they flew about the entrance of a cave at night. We collected 4 species of bats this way, but *M. keenii* was the most common. We found no bats in the cave during the day.

Distribution of *Myotis keenii* in the United States

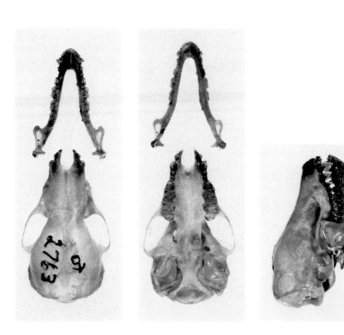

Skull of *Myotis keenii*, x 3

This species is fairly common and regularly present in caves and silica mines of southern Illinois in winter; so it is likely that it also occurs in the numerous silica mines of western Kentucky.

M. keenii is a northern species, found most frequently hibernating in caves and mines across the northern states and southern Canada. Even there, concentrations of 100 or more in one mine are unusual; this species seems nowhere abundant.

Little is known about the summer habits of this bat. It appears to be most common in Nebraska and Iowa, where it can be captured in nets set over woodland streams. Elsewhere within the range the records are primarily accidental.

These bats live singly or in small colonies of about 30 individuals or fewer. Window shutters, tar paper, and loose bark give shelter to them. Occasionally they occupy buildings, there choosing a more lighted spot than does *M. lucifugus*.

In Scioto County, Ohio, on the shore of Lake Roosevelt, just across the river from Kentucky, *M. keenii* was found living beneath the wooden shingles of a campground shelter. A few adults and their young were seen there on June 12, 1960.

This seems to be one of the scarcest bats in Kentucky. Any information that can be obtained about it would be of value.

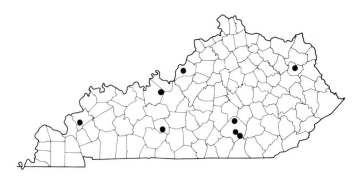

Locality records of *Myotis keenii* in Kentucky

Indiana Myotis;
Indiana Bat; Social Bat   PLATE 7

Myotis sodalis Miller and Allen

Recognition: Total length 86–94 mm; tail 35–43 mm; foot 9 mm; ear 14–15 mm; wingspread 240–267 mm; weight 5–11 g. A medium-sized, dark myotis with nonglossy fur, a slightly keeled calcar, a snub-nosed appearance, and rather small feet, the hairs on which do not extend beyond the toes.

Variation: No geographic variation is recognized in this species. Although most individuals are dark brown to nearly black, some are the same brown color commonly seen in *M. lucifugus.*

Confusing Species: Most commonly confused with *M. lucifugus*, with which it is usually found in caves in winter. Hibernating groups can often be recognized by the shape of the clusters, which tend to be tightly packed in *M. sodalis* and loose in *M. lucifugus*. However, both species sometimes hang singly. The long hairs on the feet of *M. sodalis* do not extend beyond the toes; this character alone will separate it from our other species of *Myotis* except *M. keenii* and *M. leibii*. *M. keenii* has longer ears and a pointed tragus; *M. leibii* has tiny feet and is smaller.

Kentucky Distribution: Limited, except in migration, to the region of limestone caves. The wintering colonies in Mammoth Cave National Park and Carter Caves State Park are among the largest known of this vanishing species. In recent years each contained about 100,000 bats, but the numbers are declining in spite of the protection.

Elsewhere in Kentucky we have seen small hibernating clusters in a couple of dozen caves in perhaps a dozen counties, from Bell and Elliott to Meade, Breckenridge, and Warren. West of Warren County we have not investigated the caves in winter or netted at them in summer. *M. sodalis* probably

Distribution of *Myotis sodalis*

Skull of *Myotis sodalis*, x 3

Clusters of *Myotis sodalis* hibernating in Bat Cave, Kentucky

occurs west to Trigg and Livingston counties but not in the Jackson Purchase.

Life History: This species is known in Kentucky primarily as a migrant and winter resident. Except for a few males, our populations spend the summer in Ohio, Indiana, and Michigan.

The large colony (now about 40,000) that hibernates in Bat Cave, in Carter Caves State Park, begins to disperse in late March. The females begin leaving first and head for their summer range, to the northwest. The greatest exodus occurs in late April, when nearly half the population leaves within a week. By the end of the first week of May the cave is vacant, except that a few males usually remain in the vicinity and can be netted at the cave entrance at night. Most males, however, also migrate to the northwest.

In Mammoth Cave National Park, where about 100,000 *M. sodalis* hibernate in several caves, a few hundred males spend the summer as an active band, which wanders from cave to cave.

Apparently all reproductively active females leave Kentucky to bear their young, but very little information is available about the habits of these bats on their summer range. The only record of a pregnant female was one shot as she flew along the edge of a small woodlot on June 18 in northern Indiana; she was carrying a single large embryo.

A few immature M. *sodalis* have been shot on the wing in Indiana in late July, and occasional individuals and groups of 3 or 4, including adults of both sexes and young bats, have been found under a large concrete bridge in Turkey Run State Park. A single adult was once found there beneath the loose bark of an old tree, in May.

Apparently M. *sodalis* do not use buildings as summer roosts. We know of only 2 records from buildings, both of juveniles in late summer. We found a juvenile male among more than 600 M. *lucifugus* we banded in an old building in Oakville, Logan County, Kentucky, on August 1, 1963.

Our guess is that adult females live singly or in small groups and bear their young beneath loose bark or in hollow trees. Probably a single young is born per female, in late June.

By mid-July migrant M. *sodalis* have begun to appear in Kentucky. From then on through August and September they appear in nocturnal swarms at caves. The bats appear at some caves by the hundreds; a net tended all night at the mouth of such a cave may yield nearly 1,000. The population has an almost complete turnover each day, and only rarely is an individual recaptured at the same cave. By daylight all bats disappear except in certain caves, where a few hundred sometimes rest during the day; and they are replaced by different individuals on following days.

Seven other species of bats join M. *sodalis* in swarming at Kentucky caves in the fall. We do not know the significance of the behavior nor where the bats go after leaving the caves. A female netted and banded at Dixon Cave, in Edmonson County, on September 2 was captured in a barn more than 480 km away, in St. Joseph County, Michigan, on September 10. Many of the bats appearing in August have later been found hibernating in various caves in the area where they were tagged.

When the M. *sodalis* arrive in Kentucky they are not as fat as most resident bats at this season. Their average weight at this time is a little more than 6 g. They fatten quickly in September, however, and by October most weigh 8 or 9 g.

Tightly packed hibernating cluster of *Myotis sodalis*. One bat is awake.

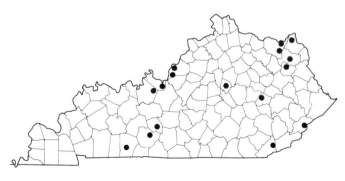

Locality records of *Myotis sodalis* in Kentucky

M. sodalis breeds in the fall. At Bat Cave most breeding occurs over a period of about 10 days in early October. At night the bats scatter in pairs over the ceiling of a large room near the upper entrance of the cave, and there hundreds can be seen copulating. By day the males cluster in the breeding room, and the females are found at the other end of the cave, about 900 m away.

As cold weather approaches, the bats gather at the hibernation sites and form the tightly packed clusters characteristic of the species. Here they pass the winter. Each individual arouses about every 8 to 10 days to spend an hour or so flying about or to join a small cluster of active bats elsewhere in the cave before returning to hibernation. Apparently even in midwinter the active bats find insects and do some feeding.

M. sodalis is on the list of rare and endangered species prepared by the U.S. Department of the Interior. It is nearly extinct over most of its former range in the northeastern states, and since 1950 the major winter colonies in caves of West Virginia, Indiana, and Illinois have disappeared. The 2 great colonies in Kentucky, both of which are protected in parks, probably offer the best hope for the survival of the species.

Small-footed Myotis; Leib's Bat   Plate 7

Myotis leibii (Audubon and Bachman)

Recognition: Total length 73–82 mm; tail 30–38 mm; foot 7–8 mm; ear 14–15 mm; wingspread 212–234 mm; weight 4–6 g. A very small, glossy-brown bat with tiny feet, a flattened skull, a keeled calcar, black ears, and a black facial mask.

Variation: There is no geographic variation in this species across Kentucky. *M. l. leibii* (Audubon and Bachman) is the subspecies occurring here. We see little individual variation among specimens we have taken in Kentucky; at most they vary slightly from lighter to darker shades of brown.

Distribution of *Myotis leibii* in the United States

Skull of *Myotis leibii*, x 3.2

Confusing Species: This tiny bat can be distinguished from all other Kentucky *Myotis* by its smaller size; it is the only one with a forearm less than 35 mm long and a foot less than 9 mm long. The strongly keeled calcar is distinctive; our only other myotis with a keel is *M. sodalis*. It is much smaller than *Nycticeius humeralis* and *Eptesicus fuscus*, both of which resemble it in general appearance. Our only bat of similar size is *Pipistrellus subflavus*, which can be distinguished by its light color—especially the light pinkish forearm—and lack of a keel.

Kentucky Distribution: Locally distributed in woodland, rocky, and cave regions of southeastern Kentucky; also known as a summer resident in Mammoth Cave National Park. Common in the western states, this species is considered one of the rarest bats in the East. Those found near Mammoth Cave represent an important major concentration of the eastern race.

Life History: In the East this species is known primarily from specimens found hibernating in mines and caves in Pennsylvania, New York, Vermont, and Ontario, where they sometimes form groups of several dozen bats. Occasional individuals are found in winter in caves as far south as Georgia. The locality nearest Kentucky where this bat is regularly found in winter is Saltpeter Cave, at Greenville, West Virginia. There are March records from a cave in Campbell County, Tennessee, just a few miles from southeastern Kentucky; this leads us to suppose that the species will eventually be found occasionally in winter in Kentucky caves.

Summer records are few and scattered. The only known colony of the eastern race consisted of about a dozen bats found behind a sliding door on a shed in Ontario. Nothing is known of the breeding habits of the eastern race.

Kentucky records of this bat are all from the warm months. An old winter record from the Mammoth Cave area was a misidentification.

In late July and in August *M. leibii* appears regularly and in fair numbers at several caves in and around Mammoth Cave

National Park. We observed this species especially when netting bats there in the summer of 1963. The small size and slow flight of the small-footed myotis made it easy to recognize among the several more common species. They seemed to detect our nets and avoid them easily, although occasionally we could chase them into a net. We found they were most easily captured in hand nets. They especially liked to get into small, dead-end passages at the entrance to caves; in these passages they could be seen fluttering about or hanging on the walls at night. On July 30, 1963, we captured 6 males and 5 females with hand nets at Dixon Cave.

Other records include a male captured in a washroom at the campground at Mammoth Cave National Park on May 9, 1952, and one found mashed on the road near the top of Big Black Mountain on July 24, 1948. In June 1973 we observed several tiny bats, which may have been this species, foraging over the road beside the radar station atop Big Black Mountain.

In June 1973 David Fassler found 2 male *M. leibii* in the expansion joints of highway bridges. One was in the bridge over Otter Creek, on Route 90 in Wayne County, the other at the Laurel–Pulaski County line where Route 192 crosses the Rockcastle River.

This is the extent of our knowledge of this species in Ken-

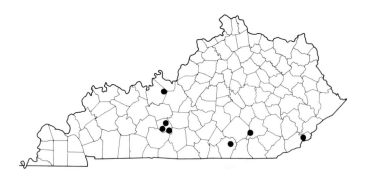

Locality records of *Myotis leibii* in Kentucky

tucky. We do not know whether it winters here; perhaps the late summer bats are only transients. Are young raised here, or does our population in May and June consist only of males? We suspect that checking bridges and open outbuildings that might serve as night roosts might give a better knowledge of the distribution and habits of this elusive bat in Kentucky.

M. leibii is one of our few bats that sometimes roosts on the ground. One was found under a rock on a Missouri hillside, and 2 were found beneath a rock in a quarry near Dayton, Tennessee. We have found them among the rocks on the floors of caves and mines in West Virginia and New York, in winter. In one West Virginia cave they spend the winter in crevices in the cracked clay floor.

Silver-haired Bat   PLATE 8

Lasionycteris noctivagans (Le Conte)

Recognition: Total length 95–108 mm; tail 37–45 mm; foot 9–10 mm; ear 14–16 mm; wingspread 270–310 mm; weight 9–15 g. A medium-sized bat with silver-tipped fur. Dorsal surface of the interfemoral membrane lightly furred; ears short, rounded, and naked; both fur and membranes black. Occasional individuals have dark-brown, yellowish-tipped fur.

Variation: No geographic variation is recognized in this species.

Confusing Species: This is a distinctive bat, which might be confused only with the much larger hoary bat, *Lasiurus cinereus*. The latter has patches of hair on the ears and is heavily furred on the upper surface of the interfemoral membrane.

Kentucky Distribution: Statewide but rare. Winter resident and migrant only; most likely to be found in April and October.

Life History: This is a northern species, characteristic of the woodland ponds and streams in the northern tier of states and

Silver-haired bat, *Lasionycteris noctivagans*, partially concealed in
a dead snag

northward in Canada nearly to the treeless zone. The southern
limits of the breeding range are poorly defined but apparently
do not include Kentucky. Occasional males linger into summer
along the east coast as far as South Carolina and down the
Appalachians at least to West Virginia. This bat is erratic in
numbers; over much of its wide range it is scarce.

The usual day roost is behind a loose piece of bark or inside
a hollow tree. In Saskatchewan a family of these bats was
found living in an abandoned woodpecker hole.

Little is known about the breeding habits of *L. noctivagans*.
Two young are produced, apparently in late June or early
July.

As fall approaches, silver-haired bats migrate southward to
their winter range, which is generally from southern Illinois
to New York City and southward nearly to the Gulf Coast.
In migration they can be encountered by day in a wide variety
of shelters. They favor open sheds, garages, and outbuildings,
rather than the enclosed attics preferred by our common brown
bats. Piles of slabs, lumber, railroad ties, and fenceposts are
also frequently chosen as temporary roosts by the migrants.

In winter, they shelter in a variety of places. Silver-haired
bats have been found hibernating beneath loose bark and in

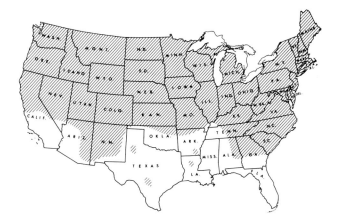

Distribution of *Lasionycteris noctivagans* in the contiguous United States

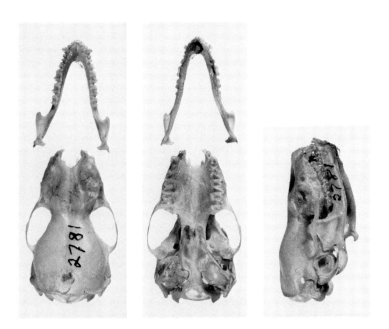

Skull of *Lasionycteris noctivagans*, x 2.6

hollow trees. A hibernating bat found in a dead tree felled in Mammoth Cave National Park in January was described to us by the park naturalist and seems to have been this species. Silver-haired bats have been found hibernating in buildings in West Virginia and in skyscrapers, churches, wharf-houses, and ships in and about New York City.

Caves are only rarely used; however, a careful observer can sometimes spot a silver-haired bat concealed in a crevice at a cave entrance. In West Virginia, Gene Frum extracted 7 from crevices in sandstone cliffs, in March. The bats were about 50 cm back in narrow (50–100 mm) cracks, where they were associated with big brown bats. They were located by probing with sticks until the squeak of a bat was elicited. The same technique would probably reveal these bats in the sandstone cliff country of the Cumberland Plateau.

Small numbers of silver-haired bats regularly winter in the silica mines of southern Illinois, where the bats are found hanging in the open on the mine wall. Thirty-five of these mines, examined in 2 winters, revealed 18 silver-haired bats. The silica mines around Salem, Kentucky, have never been examined for bats.

The best time to find silver-haired bats in Kentucky is when they begin their northward migration in spring. Persistent efforts over the latter 2 weeks of April are likely to reward a bat-hunter with a specimen. The bats appear early

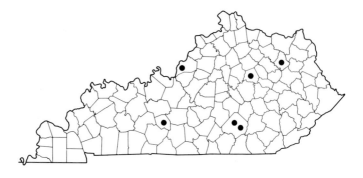

Locality records of *Lasionycteris noctivagans* in Kentucky

in the evening, and their flight is so slow that they are the easiest bats to shoot on the wing. As they are also readily captured in bat nets, netting over woodland ponds and streams may be productive at this season.

Eastern Pipistrelle   PLATE 8
Pipistrellus subflavus (F. Cuvier)

Recognition: Total length 78–94 mm; tail 35–45 mm; foot 9–11 mm; ear 13–15 mm; wingspread 208–258 mm; weight 4–7 g. A tiny, yellowish bat with tricolored fur: the base of the hairs is dark, the middle band is lighter, and the tip is dark. The anterior third of the interfemoral membrane is furred, and the calcar not keeled.

Variation: No geographic variation is evident in Kentucky. The subspecies here is *P. s. subflavus* (F. Cuvier). Kentucky specimens vary from pale yellow to orange to brown. Young are brown—much darker than adults.

Confusing Species: Similar to some of our smaller species of *Myotis* but easily distinguished from all other bats by its unique tricolored fur.

Kentucky Distribution: Statewide; abundant. May be absent from the highest elevations in the southeastern mountains.

Life History: Although it is seldom encountered by man, the little pipistrelle probably is the most abundant bat in Kentucky. Most of them probably spend the summer days hanging in the foliage near the tops of trees.

Early in the evening these dainty little bats are on the wing; they are so small and such weak, erratic fliers that they are occasionally mistaken for large moths. Some individuals fly back and forth at tree-top level, foraging over an area so small that they are constantly in view.

The pipistrelle is a forest-edge species, not found foraging in the deep woods or in open fields unless there are large trees nearby. However, it often forages over ponds and streams. In late summer, when the young are on the wing, pipistrelles are so abundant in Kentucky that we sometimes see a dozen or more in flight at one time.

Pipistrelles rarely take up residence in buildings. Occasionally, however, single transients will roost by day beneath a porch roof or in an open outbuilding, usually choosing a lighted spot in plain sight. There are a few records of small nursery colonies in buildings, but such colonies are remarkably quiet and easily overlooked; furthermore, they create neither an odor nor guano piles. A group of 2 females and 3 young we found in an abandoned house in Clark County was noted only after we had fruitlessly searched all rooms looking for bats; then one of the pipistrelles took flight, and we were able to locate the others. They were hanging from the ragged wallpaper on the ceiling in good daylight.

In Kentucky the young are born the last 2 weeks of June and the first week in July; litter size is 2. Within a month the young are flying, and from late July into August pipistrelles are at their maximum population.

In late summer these bats join other species in swarming at certain caves in Kentucky. The pipistrelles appear at the entrance, where they seem so intent on entering the cave that they give no indication of detecting a mist net, as they so often do when foraging over a pond or stream. They often hit our nets with surprising force. It is not unusual to capture 100 or more pipistrelles at a cave entrance in a single night; Wilson Baker once banded nearly 2,000 in 2 weeks of netting at Thorn Hill Cave, in Breckenridge County. This is many times the number of pipistrelles that hibernate in this cave. We wonder whence so many come and where they go.

At the approach of winter the pipistrelles retreat to the caves to hibernate. They inhabit more of our caves and mines than

Distribution of *Pipistrellus subflavus* in the United States

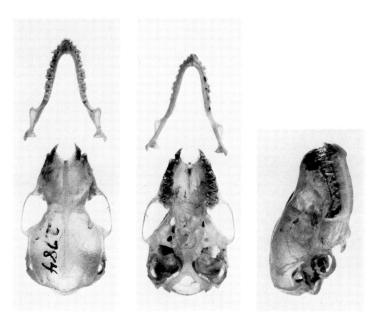

Skull of *Pipistrellus subflavus*, x 3.2

Eastern pipistrelle,
Pipistrellus subflavus

any other species of bat; still, they never assemble in large numbers. They hang singly and are scattered throughout the cave, preferring the warmer locations and the protected side-passages, where they are often the only species encountered. Their numbers in a cave range from a single individual to several hundred.

Pipistrelles choose a hibernation site where the temperature is about 11 to 13°C. An individual may occupy a precise spot in a cave in consecutive winters. It usually has several spots at which it hangs; although it shifts from one to another during the winter, it is always to be found at one of these sites. Occasionally, an individual awakens, flies about, and returns to precisely the same spot; how it can do this in total darkness is unknown.

Although pipistrelles awaken from hibernation and fly about, they do so less frequently than most other bats; an individual often remains in one position for several weeks. Beads of moisture frequently collect on the fur, giving the bat a dazzling appearance when a light is flashed on it (PLATE 8); under such conditions a careless observer frequently reports seeing a white bat. Most females leave the caves in April, but males often hibernate well into May or June.

P. subflavus is a rather hardy bat; our cave populations do not reach a maximum until well into December, and in mild winters the cave populations are smaller than in severe ones. We suspect that large numbers regularly winter outside the known caves, for our known hibernating populations cannot begin to account for the abundance of this bat in summer. The winter home of most of these delicate little bats remains a mystery.

Like our other bats, pipistrelles feed upon insects caught on the wing. In late summer they sometimes literally swarm about corncribs to feed upon emerging grain moths.

Although many thousands of these bats have been banded over the years, recoveries are scarce. The few we have suggest that this is a rather sedentary species, seldom traveling more than 80 km between its summer and winter homes.

Big Brown Bat;

Barn Bat; House Bat   P<small>LATE</small> 9

Eptesicus fuscus (Palisot de Beauvois)

Recognition: Total length 110–134 mm; tail 38–52 mm; foot 12 mm; ear 17–18 mm; wingspread 325–350 mm; weight 11–25 g (females larger than males). A large, brown bat with a broad nose, a broad, rounded tragus, broad wings, and a keeled calcar.

Variation: No geographic variation is recognized in Kentucky. Our subspecies is *E. f. fuscus* (Palisot de Beauvois). Color variation is limited, but some individuals have a russet hue and others are dark brown.

Confusing Species: Quite similar in general appearance to *Nycticeius humeralis*, which is smaller and lacks a keel on the calcar. Any large, brown bat with a keeled calcar found in Kentucky is *Eptesicus*.

Distribution of *Eptesicus fuscus* in the contiguous United States

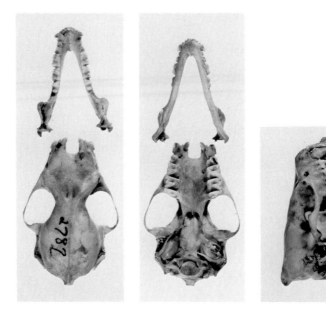

Skull of *Eptesicus fuscus,* x 2

Kentucky Distribution: Statewide; abundant. This is the commonest bat in buildings in Kentucky.

Life History: More than any other American bat, this species is associated with man. It inhabits buildings of all sorts that can provide it with adequate dark retreats and protection. At night big brown bats can be seen foraging over cities and over farm ponds or resting beneath the picnic shelters in parks.

Although they use man's structures, these bats are not entirely dependent upon us for shelter; we once found 2 big brown bats residing in summer in a crevice of a rock shelter near Old Landing, in Lee County.

In summer, nursery colonies form in attics of houses, schools, and churches, in barns, behind shutters or unused sliding doors, in the expansion joints of bridges, and in similar shelters; occasionally hollow trees are used. Colonies we have seen in Kentucky have ranged in size from a dozen adults to about 300; 50 to 100 is most common.

This is by far the most abundant bat in the city of Lexington. Along the tree-lined streets in the older residential neighborhoods we have found sections where nearly every stately brick home is used by bats. The central business district is similarly blessed. We have even found them living in modern suburban homes, in which open louvers, a dark attic, or a space between the inner walls and the brick serves their needs.

Big brown bats emerge at dusk and fly a steady, nearly straight course at a height of 5 to 10 m. The large size and steady flight makes them readily recognizable. Although at some colonies the bats may begin feeding nearby when they emerge, usually they move off in a steady stream to feeding grounds hundreds of meters away. The flight appears slow, but this is an illusion due to their large size and slow wingbeats.

After arriving at its feeding ground—a meadow, a field set off by trees, or a tree-lined city street—a bat will fly repeatedly over the same course during the evening, with frequent sallies off course to catch an insect. Apparently an individual uses the same feeding ground each night, for a bat can often be seen following an identical feeding pattern on different nights. If the bat is shot, it may be several nights before another occupies this feeding ground.

After feeding for half an hour or so, the big brown bat flies to a night roost to rest. Favored night roosts include porches on stucco or brick houses; garages with open doors; breezeways; and highway bridges. Here it leaves the telltale sign: a few droppings are found each morning. We have often been asked how to deal with such a mysterious visitor by an exasperated housewife who rarely knows what kind of animal is causing the trouble, for it is never seen. The most likely time to spot the bat is about an hour after dark. Usually such a single bat is an adult male, because the females return to their young after a feeding flight. When the young are on the wing, the group chooses the nursery site or a nearby shelter as a night roost. In Kentucky the nursery sites are used as day roosts well into October.

With the arrival of cold weather, big brown bats leave the nursery sites for their winter homes, but they are not as migratory as our other common bats. Seven recoveries of big brown bats we banded in Kentucky were at distances of 19–53 km from the banding site. The longest flight was that of a bat banded in a Bardstown colony and found in Louisville the following February. We have known several individuals to spend winter and summer in the same locality.

Big brown bats hibernate in caves, mines, storm sewers, and buildings. They are very hardy, often choosing sites where the temperature drops below freezing. In caves and mines they use sites near the entrance, where they may come and go even in cold weather.

We have not been able to determine where the major population of this very common bat is to be found in winter. To find more than 25 or so in a Kentucky cave is unusual, and all the winter groups we have seen in buildings have been even smaller.

Breeding takes place in the fall and winter, and ovulation occurs about the first week in April. Maternity colonies begin

Baby big brown bats, *Eptesicus fuscus*, clustered in the attic of a building while their mothers are foraging. Two adults have already returned and are nursing their young.

to form in the latter part of April, and most of the females have arrived by the middle of May. Some colonies contain only females; in others a few males are found hanging singly, away from the clusters of females. As the young mature, there is an influx of adult males, and for the rest of the season they may be nearly as numerous as adult females.

The young are born in the last week in May and the first week in June; 2 is the normal size of the litter. The young are large at birth, weighing about 3 g. They are left in a cluster when the mothers go out to feed, and, upon returning, each mother selects her own young to nurse. A young bat that loses its grip and falls to the floor is commonly retrieved by its mother, although many such youngsters perish. A mother whose babies have been taken away from her will visit squeaking youngsters on the floor at dusk, but she will retrieve one of them only if it is her own.

The young grow rapidly. Increases in forearm length of as much as 2.6 mm were recorded in a single day, and weight gains of 0.5 g per day are common among the newborn. When the young are small the mothers forage close to the colony and return to nurse within an hour; as the young mature they are left for longer periods. By mid-June the mothers begin to use night roosts away from the nursery, and about a week later the juveniles are on the wing and join them at the night roosts. By early July all the young are able to fly.

Although colonies are usually found in hot attics, this species

Louver in the belfry of a church in Clay City, Kentucky, occupied in summer by a colony of big brown bats, *Eptesicus fuscus*. The stain at the top indicates where the bats go in and out. Such stained areas are often rather conspicuous and betray the usage of a building by bats.

is not nearly so tolerant of heat as is the little brown bat. When the temperature in a colony rises to 33 to 35°C the bats move down from the roof of an attic onto the floor or into the walls. During hot spells in the latter half of June young bats that are unable to fly may find their way into living quarters and various other places of contact with man. At such times we always receive a flurry of calls about bat problems in Lexington.

In spite of the fact that *E. fuscus* is a rather sedentary species, it is one of the most accomplished at homing. Individuals have returned to Cincinnati after having been conveyed 725 km to the north. At a colony in Indiana nearly all bats that were displaced 402 km returned on the fourth and fifth nights.

Like our other bats, this species is generally beneficial to man. Its diet consists entirely of flying insects—mostly beetles, members of the wasp group, and flies. Colonies that invade buildings sometimes become a genuine nuisance. In houses, churches, and schools they can generally be eliminated by closing their entrance holes after they leave in the fall. However, we have seen barns for which we could devise no way of eliminating the bats.

This is the only Kentucky bat that we have been able to get to adapt well to captivity. Although they are vicious at first and can inflict a painful bite, they quickly become gentle.

They soon learn to eat a mixture of chopped insects, canned dog food, cottage cheese, and banana, and some individuals learn to eat just dog food.

Red Bat PLATE 9

Lasiurus borealis (Müller)

Recognition: Total length 96–120 mm; tail 40–56 mm; foot 8–10 mm; ear 11–13 mm; wingspread 290–332 mm; weight 8–16 g. A medium-sized, red or reddish bat with long, pointed wings, short, rounded ears, and a heavily furred interfemoral membrane. When the bat is in flight its long tail extends straight out; thus this bat can be recognized, when foraging with our other common bats, by its distinctive silhouette.

Variation: No geographic variation is evident in Kentucky. The subspecies here is *L. b. borealis* (Müller). Color varies from bright red-orange through yellowish to nearly brown; males are brighter than females.

Confusing Species: No bat known from Kentucky could be confused with this species. *L. seminolus*, a southern species that wanders northward in August and may someday be taken in Kentucky, is a deep mahogany color but otherwise appears identical to the red bat.

Kentucky Distribution: Statewide; abundant.

Life History: Among the wooded hills of eastern Kentucky red bats appear on the wing very early in the evening. They fly so high above the valleys that one strains to see them and recognize that they are bats. The bats continue in erratic flight—perhaps in play, perhaps catching insects—for half an hour or more, until dusk brings out our other species of bats. At this time red bats descend to within a meter or so of the

Family of red bats, *Lasiurus borealis*, in their daytime shelter beneath the leaf of a sycamore tree. The two young are nearly as large as the mother.

ground to feed, and they also drink from ponds or streams. The flight pattern changes at dusk from slow and erratic to swift and straight.

The red bat is a tree-dweller. Although abundant, it is rarely seen by man, for it seldom enters buildings. Only in late June, when females nursing young sometimes fall onto suburban lawns, do people and red bats tend to meet. It is unfortunate that so few people see this species, because its bright reddish color and silky fur make the male red bat one of the most attractive of Kentucky mammals.

For its day roost the red bat chooses a dense cluster of leaves. Hanging by one foot from a twig or the petiole of a leaf, the bat is surprisingly well concealed: it closely resembles a dead leaf. The bats prefer trees in a fencerow, at the forest edge, or among suburban homes. They generally choose the south side of a tree and select a spot 1 to 6 m above ground. The roosting site usually provides protection from view from any direction except from below; it lacks obstruction beneath, perhaps so that the bat can freely drop downward to begin flight. The patient observer can learn to spot red bats by seeking their roosts. A good roost site is often used by different individuals on different days.

Although they are not cave-dwellers, red bats regularly join other species in swarming about certain Kentucky caves in August and September. They fly in and out of the caves at night and are easily captured in mist nets. We often take a dozen or so in an evening's netting.

Red bats sometimes fly deep into the caves. In places where there are small passages leading to large rooms, they may get lost in a room and come to rest on the ceiling. Then they go into a torpid state, from which they cannot arouse spontaneously, because of the constant low temperature; so, eventually, they die. There is a room in Bat Cave, in Carter Caves State Park, where we can usually find 3 or 4 such red bats in August.

Red bats are migrants, moving southward in late summer and fall. One we banded at Mammoth Cave in August was later found dead in Texas—as far as we know, the only recovery of a banded red bat. Knowledge of the migration patterns awaits more extensive banding programs.

Red bats are abundant in Kentucky in winter, although their exact whereabouts remains a mystery. On any winter day when the temperature rises above 18°C red bats can be seen flying and catching insects in late afternoon. In January 1957 we counted 27 as we drove along the woodland roads in Mammoth Cave National Park. Red bats apparently pass the winter hibernating in sheltered sites in trees, but all efforts to locate one have been unsuccessful.

As with most of our bats, breeding occurs in fall. Copulation is initiated in flight, and usually the breeding pair falls to the ground, but there is one observation on record of a pair mating while flying.

Fertilization occurs in spring after the bats emerge from hibernation. The young are born in late May or early June. Red bats produce the largest litter of any bat in Kentucky; 3 or 4 young are common. Mortality of the young probably is rather high. Blue jays prey upon them persistently and probably take a heavy toll. Surviving young are on the wing 3 or 4 weeks after birth.

Little is known about the food of red bats. They are known

Distribution of *Lasiurus borealis* in the United States

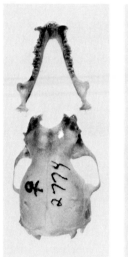

Skull of *Lasiurus borealis*, x 2.5

to eat crickets, flies, bugs, beetles, cicadas, and moths. The presence of crickets suggest they must take some food from the ground. They commonly feed beneath streetlights in towns and occasionally can be seen to alight and capture an insect on a wooden light pole. Perhaps more than any other bat, this species takes advantage of the attraction of insects to lights. Black-light insect traps attract and occasionally capture red bats. We have obtained more than a dozen specimens from such traps at the University of Kentucky experimental farms.

Another accident that commonly befalls the red bat is impalement upon a barb of the top strand of a barbed-wire fence. Apparently the bat mistakes the barb for an insect and punctures its wing or tail membrane in trying to capture the "prey."

Hoary Bat PLATE 10

Lasiurus cinereus (Palisot de Beauvois)

Recognition: Total length 134–148 mm; tail 53–61 mm; foot 12–14 mm; ear 18–20 mm; wingspread 380–410 mm; weight 28–35 g. A large, dark-colored, heavily furred bat. Dorsal surface of interfemoral membrane completely furred; ears relatively short, rounded, and edged with black. The tips of many hairs are white, giving the bat a frosted, hoary appearance. This is the largest bat in Kentucky.

Variation: No geographic variation is evident in Kentucky. The subspecies here is *L. c. cinereus* (Palisot de Beauvois).

Confusing Species: The only other bat resembling the hoary is the much smaller silver-haired bat, *Lasionycteris noctivagans*.

Kentucky Distribution: Statewide; rare, but likely to be present at all seasons. Probably more common as one goes westward in Kentucky.

Distribution of *Lasiurus cinereus* in the contiguous United States

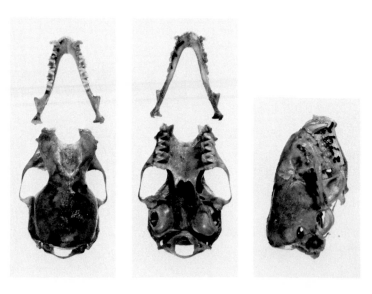

Skull of *Lasiurus cinereus*, x 2.1

Characteristic defensive
pose of a hoary bat,
Lasiurus cinereus

Life History: The hoary bat spends the summer days concealed in the foliage of a tree. A leafy site, well covered from above but open beneath and 3 to 5 m above ground, is generally chosen. In Kentucky our sparse summer population of hoary bats consists entirely of adult females and their young. Juvenile males appear in fall migration; in summer the adult males occur only in western North America.

Long after other bats are on the wing and darkness has settled in, the hoary bat launches into flight. The flight is swift and more direct than that of our other bats, and the flight pattern and large size make the hoary bat readily identifiable on the wing.

In migration hoary bats seem to be a little more common in Kentucky than at other seasons. At these times they fly earlier in the evening and are more likely to be seen. We have found them in April, May, September, October, and November.

This is one of the easiest bats to capture in a mist net. The most likely way to see one in Kentucky is to set a net across a pond and watch it for 2 or 3 hours after darkness has fallen.

Litter size is 2, and the young are generally born in June. When the young are half-grown or larger, the mother and young are sometimes dislodged from their perch and fall to the ground. Because the mother is unable to fly with such a load, she and her babies may lie helpless and come to the attention of man. Our only summer records of hoary bats in Kentucky are of families found this way in Lexington and brought to our attention.

In western Kentucky the hoary bat is probably not as scarce as records would indicate. Extensive netting over ponds in the Purchase in the migration period would probably reveal a considerable number of migrants.

Evening Bat   PLATE 10
Nycticeius humeralis (Rafinesque)

Recognition: Total length 94–105 mm; tail 36–42 mm; foot 9 mm; ear 14–15 mm; wingspread 260–280 mm; weight 8–14 g. A small to medium-sized, brown bat, lacking distinctive external features. Calcar not keeled; tragus short, curved, and rounded. The skull is broad, especially anteriorly, and there are but 2 upper incisors.

Variation: No geographic variation is recognized in Kentucky. The subspecies found here is *N. h. humeralis* (Rafinesque).

Confusing Species: This species is frequently misidentified. It is often confused with *Myotis* species, even though it is easily separated by its short, curved, rounded tragus. From the big brown bat, *Eptesicus fuscus,* it can be separated by its smaller size and its lack of a keel on the calcar.

Kentucky Distribution: Western and southern Kentucky. Uncommon in summer; apparently absent in winter.

Life History: This is a colonial species, inhabiting buildings and cavities of trees. As cold weather approaches, the Ken-

Distribution of *Nycticeius humeralis* in the United States

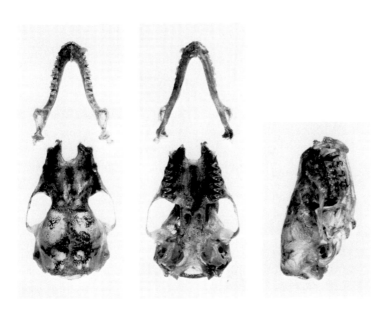

Skull of *Nycticeius humeralis*, x 2.6

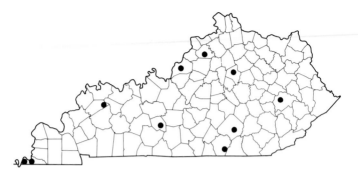

Locality records of *Nycticeius humeralis* in Kentucky

tucky population disappears, perhaps by migration southward. This is the only species of bat in Kentucky that we have never found in a cave; however, it does occasionally join cave species in late-summer swarming activities. We have captured half a dozen among the thousands of other bats netted over a trail just outside Dixon Cave, in Mammoth Cave National Park.

Maternity colonies consisting of 100 bats or more have been found in houses and other buildings at several sites in Indiana and Illinois. In Kentucky we have heard of but one such colony—at Onton, in Webster County; but in our studies of colonial bats we have failed to locate a colony of this species.

N. humeralis emerges early in the evening and flies a slow, steady course. It is readily recognizable in flight by the experienced observer.

Litter size is 2. Judging by data from Illinois and Indiana, most young in Kentucky are probably born in the first week of June.

The young are pink and naked at birth. Within 24 hours the eyes open and heavy skin-pigmentation appears, and soon thereafter the ears unfold and become erect. In the third week of life the young begin to fly, and at the end of the third week they can negotiate turns and can land on the walls and ceiling. They are weaned at 6 to 9 weeks of age.

This is one of the least-known bats in Kentucky. Although it is a southern species, nearing the limits of its range in

Indiana and Illinois, it is definitely more common in those states than in Kentucky. We have seen only one individual in the Lexington area and believe that the species is quite scarce in the Bluegrass. David Fassler, who has done extensive collecting about Somerset in recent years, has taken only one evening bat.

Townsend's Big-eared Bat; Western Big-eared Bat PLATE 10

Plecotus townsendii (Cooper)

Recognition: Total length 101–116 mm; tail 40–54 mm; foot 10–13 mm; ear 33–38 mm; wingspread 297–324 mm; weight 7–12 g. A medium-sized, brown bat with large ears. Two large lumps appear on the dorsolateral surface of the snout.

Variation: Several subspecies are known, but no geographic variation occurs in Kentucky. Our subspecies is *P. t. virginianus* (Handley). Color in our population varies from pale to dark brown; the underparts are light brown.

Confusing Species: Our only other big-eared bat is *P. rafinesquii,* a species easily recognized by its white belly, long hairs on the toes, and bicuspid upper incisors.

Kentucky Distribution: Rare; known only from a single colony in a cave in Lee County and from individuals that apparently wandered from this colony into many caves in the area, including those of 4 neighboring counties. This is a vanishing race of a western species.

Life History: This species is represented in the East only by a few relict populations inhabiting limestone caves. It was first recorded in Kentucky on October 12, 1952, when 2 were found

Distribution of *Plecotus townsendii* in the United States

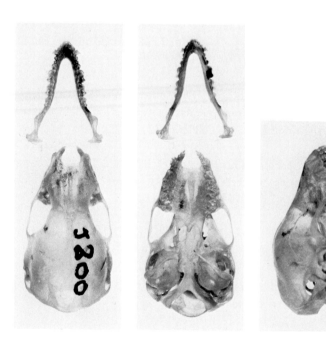

Skull of *Plecotus townsendii*, x 2.8

Two Townsend's big-eared bats, *Plecotus townsendii*, hibernating in a cave in Lee County

hibernating in a small cave in Natural Bridge State Park, in Powell County. The first awareness that a substantial colony of this rare bat exists in Kentucky came with the discovery of a nursery group in a cave in a remote part of Lee County in June 1963. The following winter about 1,000 of these bats were found hibernating in the cave.

In the East this is strictly a cave bat. The only records that are not from caves are of individuals apparently in transit. Often the same cave is used both winter and summer, as is the case with the Kentucky colony.

This is one of the most secretive of bats, and little is known of its behavior. It does not emerge until dark. It is most intolerant of human disturbance and thus is one of our most endangered species of bats, leaving its home in the caves as more people venture therein.

The bats form tight clusters on the walls and ceilings of a cave. In summer the animals are alert and take flight at the least disturbance; a light shone on the group not only will put it to flight but is likely to cause the group to abandon the site for days or months.

Summer colonies consist entirely of adult females and their young; the whereabouts of the adult males at this time is unknown. In West Virginia a cluster of males was found in a cave in August.

The single young is born in June. It is a grotesque creature, with its large ears flopping over the unopened eyes. It grows rapidly and takes flight when about 3 weeks old. When a month old it is nearly full-grown.

As with many other species of bats, the nursery colonies apparently break up in August, after the young are weaned. Only 40 bats were counted in the cave in August 1963, although a nursery colony of about 300 had been present in June. In October about 1,000 were present. A count made on October 28, 1964, when the bats were banded, revealed that exactly half of the 850 were males—an indication that the adult males were in the hibernating group.

Hibernating *P. townsendii* are not much more tolerant of human disturbance than are the nursery colonies. Many leave the caves in which they are banded and may later be found hibernating in other caves.

Of the many caves in Lee County that occasionally harbor big-eared bats, we have explored several in which we found both species—*P. townsendii* and *P. rafinesquii*—at the same time.

Winged parasitic flies of the family Streblidae are common on these bats. These rather large, yellow insects are quite conspicuous and can be found on the bats at all seasons. They

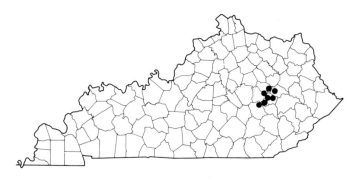

Locality records of *Plecotus townsendii* in Kentucky

are abundant; half a dozen can sometimes be found on a single bat.

This is a gentle bat to handle. Only occasional individuals attempt to bite, and they are never as vicious as many other species. However, they adapt poorly to captivity. Intensive efforts by people interested in studying their echolocation system have not been successful in getting these bats to survive for more than a few days in the laboratory.

Big-eared bats are among the most interesting of all our mammals. Because they are so rare, so intolerant of disturbance, and of such limited distribution in Kentucky, every effort should be made to protect the caves in which they live.

Rafinesque's Big-eared Bat;
Eastern Big-eared Bat   PLATE 11
Plecotus rafinesquii Lesson

Recognition: Total length 94–110 mm; tail 43–54 mm; foot 10–13 mm; ear 29–37 mm; wingspread 265–301 mm; weight 10 g. A medium-sized, light-bellied, gray bat with the basal portion of the hairs black, contrasting sharply with the tips. Long hairs extend well beyond the toes. The ears are large and there are 2 large lumps on the dorsolateral surface of the snout. The first upper incisors are bicuspid in nearly all individuals.

Variation: No geographic variation is evident in Kentucky. The subspecies here is *P. r. rafinesquii* Lesson.

Confusing Species: The only other big-eared bat in Kentucky is *P. townsendii*, which can be recognized by its tan belly, short hairs on the toes, and unicuspid upper incisors.

Kentucky Distribution: Statewide except the Bluegrass; uncommon.

Distribution of *Plecotus rafinesquii*

Skull of *Plecotus rafinesquii*, x 2.6

Rafinesque's big-eared bat, *Plecotus rafinesquii*

Life History: In Kentucky this big-eared bat inhabits caves and buildings. Two caves, in Wayne County and Jackson County, each harbor colonies of about 100 bats—the largest known colonies of this species. The bats are present year-round. In Mammoth Cave National Park a small cave harbors a nursery colony of about 50 bats, and an abandoned building contains a similar colony. The whereabouts of these bats in winter is unknown.

This species always emerges from its roosts after the last light of evening is gone. At Robinson Forest, in Breathitt County, where a colony occupies several buildings, we have often shot bats in the evening as they flew along the narrow valley, but never a *Plecotus*. We once took one in a net set over the stream after dark, however.

When approached by day in summer, these bats are immediately alerted and begin waving their ears, apparently in an effort to keep track of the intruder by means of echoes from him or by hearing the sounds he produces. The bats also give the impression that they are peering intently at the

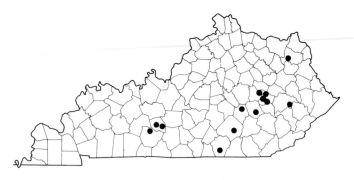

Locality records of *Plecotus rafinesquii* in Kentucky

observer. If put to flight they are remarkably agile and difficult to catch, even when pursued with a net in a small room.

Breeding apparently occurs in fall or winter. The single young is born in the latter part of May or first half of June.

This is one of the least studied of American bats. Although scarce throughout its range, *P. rafinesquii* are perhaps as common in Kentucky as anywhere. Look for them in old abandoned buildings, where they may be found in the open in plain sight, often in well-lighted areas. They are easily overlooked and leave scant sign of their presence; usually they are first seen when they take flight.

You may search dozens or even hundreds of Kentucky caves without finding this bat. The caves of Lee County are the only ones in which we have found it regularly, and there have been only a few individuals at a time. Its patterns of distribution and population numbers are mysterious.

This bat may winter in the silica mines of western Kentucky, north of Lake Barkley, inasmuch as the species occurs regularly in small numbers in winter in the silica mines of southern Illinois.

At Reelfoot Lake, Tennessee, just across the border from the western tip of Kentucky, a colony of *P. rafinesquii* inhabits an abandoned cistern. The population fluctuated from 1 to 64 in the course of fall, winter, and spring observations over a period of several years.

Family Molossidae Free-tailed Bats

Members of this family have the tail extending well beyond the rear edge of the tail membrane. They all have short, dense, dark-brown to black fur, and they give off a musty odor. They are colonial, primarily in caves but also in buildings. Sometimes the colonies contain millions of individuals. In Kentucky this family is represented by only one species.

Brazilian Free-tailed Bat; Free-tailed Bat; Guano Bat   Plate 11

Tadarida brasiliensis (I. Geoffroy St.-Hilaire)

Recognition: Total length 93–102 mm; tail 30–37 mm; foot 10–12 mm; ear 19–20 mm; wingspread 290–325 mm; weight 8–14 g. A rather small, dark-brown bat with long, narrow wings and deep vertical grooves on the upper lip. About half of the tail extends free of the interfemoral membrane. The ears

Brazilian free-tailed bat, *Tadarida brasiliensis*

● Record of autumn wanderer

Distribution of *Tadarida brasiliensis* in the United States

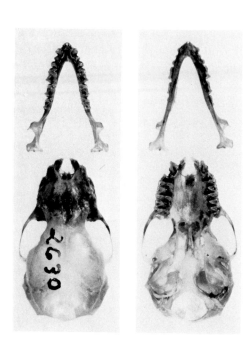

Skull of *Tadarida brasiliensis*, x 2.6

Foot of *Tadarida brasiliensis*. The long stiff hairs are distinctive.

almost meet at the midline but are not joined. Hairs as long as the foot protrude from the toes.

Variation: No geographic variation is evident in Kentucky. The subspecies here is *T. b. cynocephala* (Le Conte).

Confusing Species: This bat is unique among Kentucky mammals. No other free-tailed bat is likely to be found in the state.

Kentucky Distribution: Accidental wanderer. There is one recent record from Murray, in Calloway County, and there is a Pleistocene deposit of a colony in Mammoth Cave. A record from near Portsmouth, Ohio, across the river from northeastern Kentucky, suggests that this species may occur as an occasional straggler anywhere in the state.

Life History: This is a tropical species, ranging from South America into the southern United States. Throughout most of its range it inhabits caves as well as buildings; this is the bat that forms nursery colonies numbering into the millions in caves in the Southwest. However, the southeastern U.S. race, to which our specimen belongs, seems to avoid caves.

The breeding range of this species in the Deep South extends as far north as the southern half of the Gulf states. The northernmost breeding colonies that we know of are at Tuscaloosa, Alabama, and Forsythe, Georgia.

Nursery colonies that form in buildings in the southern states commonly contain many thousands. Breeding occurs in February and March, and, following a gestation period of 77 to 84 days, the single young is born in June. Growth is rather slow, compared with that of other bats: the young first take flight at about 5 weeks of age.

As fall approaches, the great colonies break up, and in September and October the bats, especially the young of the year, may wander hundreds of miles beyond the breeding range. It is at this time of year that stragglers occasionally are found as far north as Ohio, Illinois, and Kentucky.

The winter habits of our race are poorly known. A few individuals remain at some of the nursery colonies through the winter, but most of the colonies are deserted, even in Florida. In cold weather these bats can hibernate for short periods.

During our fall experiences of netting many thousands of transient bats at the entrances of certain Kentucky caves, we have always been on the lookout for this species, which is easily captured in nets and frequently associates with other bats in areas where it is common. That we have never captured one here suggests that this bat is indeed a very rare straggler this far north.

ORDER LAGOMORPHA *Rabbits*

The lagomorphs differ from rodents in many respects, but the easiest way to distinguish them is by the presence of 4 upper incisors; there is a second, peglike pair immediately behind the front upper pair. The maxillary tooth-rows are farther apart than the mandibular rows, and only one pair of rows is capable of opposition at a time. The facial portion of the maxillary bone is incomplete. The body is well furred, and the tail is very short.

The order is native to all of the world's major land masses except Australia and southern South America, and it has been introduced into both these areas. In Kentucky the order is represented by 3 species of *Sylvilagus*, of the family Leporidae.

1. a. Total length about 535 mm; weight about 1.3–2.7 kg: *Sylvilagus aquaticus*, Swamp Rabbit, p. 129
   b. Total length less than 490 mm; weight about 0.9–1.6 kg: 2

2. a. Total length about 390 mm. Ears small, with a black patch between them. Supraorbital very small; posterior process short, tapering posteriorly to a slender point, free from or barely touching skull, and narrowing anteriorly until anterior process and notch is usually absent: *Sylvilagus transitionalis*, New England Cottontail, p. 127
   b. Total length usually more than 400 mm. Ears larger, generally with no black spot between them. Supraorbital broadly developed; posterior process usually broadly strap-shaped and coalescing with the skull posteriorly and sometimes along its entire length; anterior process broad and commonly extended to the nearly closed anterior notch: *Sylvilagus floridanus*, Eastern Cottontail, p. 121

2a 2b

Eastern Cottontail Plate 12

Sylvilagus floridanus (Allen)

Recognition: Total length 375–490 mm; tail 39–70 mm; hind foot 80–108 mm; ear from notch 53–66 mm; weight 0.9–1.6 kg. A small rabbit, generally of rusty, buffy-brown appearance but varying from a rather dark gray to a reddish color.

Distribution of *Sylvilagus floridanus* in the United States

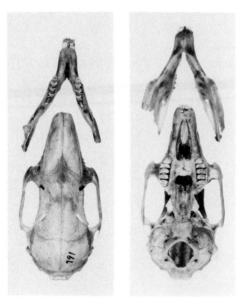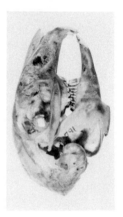

Skull of *Sylvilagus floridanus*, x 0.4

Variation: The only subspecies recognized in Kentucky is *S. f. mearnsii* (Allen). However, our state seems to be a region of intergradation of three races. Specimens from southern Kentucky might be expected to have the somewhat longer ears and shorter hind feet characteristic of *S. f. mallurus* (Thomas). Since specimens from western Tennessee and southern Illinois have been referred to *S. f. alacer* (Bangs), we might expect to find this race in the Purchase. It is a small race, with the median dorsal surface a deep ochraceous buff, strongly washed with black. Introductions of various subspecies for purposes of restocking have confused the situation in Kentucky.

Confusing Species: The swamp rabbit, *S. aquaticus,* of western Kentucky is much larger, with short, sleek fur and a thinly haired tail. *S. transitionalis* is very similar to *S. floridanus* but is somewhat smaller, with shorter ears, darker pelage, and a black patch between the ears. These 2 species are best distinguished by skull characteristics (see key).

Kentucky Distribution: Statewide; abundant. This is the common rabbit in Kentucky.

Life History: The eastern cottontail occupies a wide variety of habitats. It is found on lawns and in parks even in our largest cities, where it holds its own in spite of heavy losses to automobile traffic and to raids by cats and dogs upon the helpless young in the nest.

Although the species is most common in upland thickets and brushy farmland, individuals are occasionally found in the deep woods and lowland swamps. Nearly any habitat in Kentucky supplying some food and cover will have a few cottontails.

Most of the day, a rabbit sits quietly in a form, which is sometimes well protected in a brush pile or a blackberry thicket but is often made in the scant shelter of a tussock of grass, a flower bed, or a grown-up asparagus patch. The animal is easily overlooked, because it will sit tight unless you approach to within a meter or so. Unless badly frightened it is likely to

Young eastern cottontail, *Sylvilagus floridanus*

return to the same spot. Although rabbits are active mostly at night, they can frequently be seen feeding in the evening and early morning.

After a snow the distinctive tracks of the cottontail betray the course of his activities during the previous night. Tracks of one rabbit are often so numerous as to give a false impression of the abundance of these animals.

Rabbits feed upon a wide variety of green vegetation. On lawns they seem to prefer clover to grass. In gardens they are fond of newly sprouted peas and beans and can sometimes become a nuisance. Their winter food consists of what suitable greenery they can find, as well as fallen apples, blackberry canes, and bark. When heavy snow covers the ground, they feed upon the bark of young trees; at such times they can do severe damage to a new apple orchard, by girdling and killing trees.

Rabbits are prolific—proverbially and in fact. Litter size ranges from 3 to 8, averaging about 4 or 5, and the gestation period is about 30 days. The first litter appears in March; the doe is bred again immediately; and litters follow one another every month through September. By late summer the early litters are producing young of their own.

The young are born in a nest, which the mother makes

by scraping out a shallow depression in the ground. Although such a nest is commonly made in a flowerbed or even on a well-tended lawn, it is remarkably well concealed and seldom found. The nest is lined with finely shredded grass or leaves and quite a bit of rabbit fur, which the mother pulls from her belly and breast. Young rabbits are blind, naked, and helpless at birth. They grow rapidly and leave the nest to shift for themselves within 3 weeks.

The extent of the home range of a rabbit varies considerably: from less than 0.4 hectare to as much as 4 hectares, depending upon the supply of food and cover. Like other animals, rabbits are reluctant to leave the home range. The hunter takes advantage of this fact; he knows that a rabbit, when chased by dogs, will circle back time and again across its range, unless it is able to escape into a hole.

Rabbits have many enemies. Although they live for 10 years or so in captivity, few in the wild survive for more than a year. About 35% of the young die in their first month, and the annual mortality among the survivors is at least 65%. Dogs, cats, foxes, skunks, crows, snakes, and other predators raid the nests. Heavy rains often drown the young. Older rabbits are eaten by hawks, owls, mink, bobcats, foxes, and others. Highway traffic takes a heavy toll, as does the hunter. In spite of

Rabbit tracks in the snow

all this, rabbit populations remain high, because the animals are so prolific. The main factor limiting rabbit numbers is the supply of food and, especially, of cover. Rabbit populations are low on well-kept farms that have no brushy fencerows and in forests where a ground cover of leafy plants is nearly absent.

Remarks: As one of Kentucky's major game animals, the eastern cottontail is important to hunters, who show great concern about its numbers. As with other species of wildlife, the population varies widely from year to year. When the population is low or when the hunter fails to find as many rabbits as usual at his favorite spot, there is an outcry about the disappearance of the rabbit. Many blame predators and want a campaign against foxes; others advocate a stocking program. Both types of efforts are futile. The rabbit population can most easily be increased by replenishing the vegetation they need for food and cover and by allowing a few brush piles to remain and fencerows to grow up.

When you find fewer rabbits at your favorite spot, there are also some other factors to consider. As a result of ecological succession the rabbit's home changes with time, and not always for the better. As optimal habitat ages and the trees come in, it becomes less desirable for rabbits each year and will support fewer of them. On the other hand, an old farm that had inadequate cover a few years ago may be optimal now. You may simply need to find a new place to hunt.

Your own age may be a factor. Remember how many rabbits you used to kick up on that hillside? You were younger then, and perhaps more active in covering the ground.

Actually, the population of rabbits in Kentucky is much higher now than it was before the white man came. Because rabbits prefer second-growth vegetation to mature forest, the clearing of the land for agriculture and settlement increased their population. With the trend toward abandonment of farms and state purchase of wildlife-management public hunting areas, the future of the cottontail in Kentucky is good. There are factors, however, that are working against the hunter.

Rabbit track; the animal was traveling from left to right. The prints of the hind feet are in front of those of the forefeet.

As the human population expands, more land is used for housing, shopping centers, airports, highways, and industrial sites. Although some of these land uses provide adequate habitat for rabbits, hunting is outlawed in all of them.

Rabbits rank second to the gray squirrel in the number of game mammals harvested in Kentucky and also in the number of hunting trips for specific targets. Estimates by the Kentucky Department of Fish and Wildlife Resources are that an average of 950,000 rabbits of all species were taken annually in Kentucky in the 1964–65 through 1970–71 hunting seasons.

New England Cottontail   PLATE 12

Sylvilagus transitionalis (Bangs)

Recognition: Total length 355–450 mm; tail 35–57 mm; hind foot 81–110 mm; ear from notch 48–60 mm; weight 0.9–1.6 kg. A dark rabbit, slightly smaller than the common cottontail, with a black spot between the ears. The ears have a distinct band of black or dark brown along the anterior outer edge.

Confusing Species: The common eastern cottontail, *S. floridanus*, is very similar. It has proportionately longer and narrower ears, lacks the black spot between the ears, and has certain distinctive skull characteristics. In Connecticut the 2 forms intergrade and cannot be distinguished, but they seem to behave as separate species elsewhere.

Distribution of *Sylvilagus transitionalis*

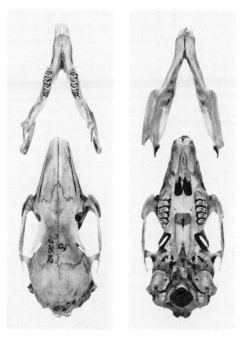

Skull of *Sylvilagus transitionalis*, x 0.7

Kentucky Distribution: Known only from a single specimen taken at an elevation of 1,190 m on Big Black Mountain in Harlan County (the highest mountain in Kentucky). The species may occur on some of the other high points and ridges in southeastern Kentucky.

Life History: Little is known about the life history of the New England cottontail anywhere, and essentially nothing is known about it in Kentucky. It is more of a forest animal than the eastern cottontail, its favored habitat being woodland brush and small clearings in the forest. The species is rarely seen, apparently because it is somewhat more nocturnal than our common rabbit. A good place to see one is in the clearing beneath a fire tower in the high Appalachians, at daybreak.

Several litters are born each year, from late winter until mid-September.

Swamp Rabbit   PLATE 13
Sylvilagus aquaticus (Bachman)

Recognition: Total length 520–540 mm; tail 67–71 mm; hind foot 105–110 mm; weight 1.3–2.7 kg. This is a large, big-headed rabbit with short fur and a slender, thinly-haired tail. Color grayish-brown above, washed with black; underparts white except for the buffy underside of the neck; front legs and tops of hind feet reddish-brown.

Variation: No geographic variation is evident in Kentucky. Our subspecies is *S. a. aquaticus* (Bachman).

Confusing Species: The eastern cottontail, *S. floridanus*, is the only similar species within the range of the swamp rabbit in Kentucky. It can be distinguished by its smaller size, longer fur, and powderpuff tail.

Distribution of *Sylvilagus aquaticus*

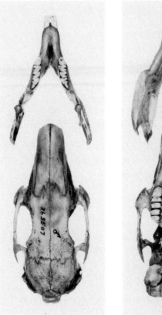

Skull of *Sylvilagus aquaticus*, x 0.5

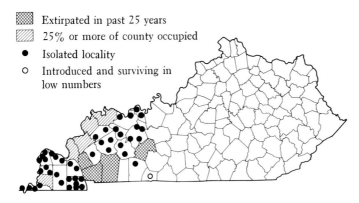

Extirpated in past 25 years
25% or more of county occupied
Isolated locality
Introduced and surviving in
low numbers

Distribution of *Sylvilagus aquaticus* in Kentucky

Kentucky Distribution: Western Kentucky eastward to Hancock, Ohio, and Logan counties. The range is shrinking rapidly as the wetlands are drained. The species has already been extirpated in Butler, Christian, Todd, and Lyon counties. It is common only in some of the counties bordering the Ohio and Mississippi rivers.

Efforts have been made by the Department of Fish and Wildlife Resources to establish this fine game species in suitable habitat in half a dozen counties in central and southern Kentucky, but apparently none has succeeded.

Life History: This is an animal of the lowland swamps and wooded floodplains. It is well adapted to a semiaquatic life. Populations on the numerous islands in the Ohio River ride out the floods by clinging to trees.

The swamp rabbit is an important game species in 15 western counties, and in 4 of these it is still an abundant animal. Populations as high as 1.2 rabbits per hectare have been recorded on some islands. However, the trend is downward, and the future of the swamp rabbit in Kentucky is in jeopardy because of habitat destruction.

Timber is being cut on one-fourth of the swamp rabbit habitat in the 15 most important counties. Oil pollution and coal mining have destroyed other swamps formerly inhabited

Swamp rabbit, *Sylvilagus aquaticus* and woodchuck, *Marmota monax*, riding out a flood on the Ohio River

by these rabbits. The greatest threat to the species in Kentucky, however, is drainage of the swamps by the U.S. Army Corps of Engineers and the U.S. Soil Conservation Service. As the swamps disappear, so do the swamp rabbits and the other animals adapted to live in the swamplands.

Two to 5 litters are produced per year. Litter size is from 2 to 6. After a gestation period of 40 days the young are born in a shallow, surface nest lined with shredded grass and rabbit fur. The nest is sometimes made in a hollow log or stump. At birth the young are well furred, but their eyes and ears are closed. The eyes open and the young begin to walk about in 2 or 3 days.

Remarks: Unlike our other rabbits, a "swamper" will often elect to spend the daylight hours in a dry spot that it can reach only by wading or swimming. It is always surprising to us to jump a rabbit from a log or stump projecting out of shallow water and have the animal go bounding away, creating a great splash every time it strikes the water.

Juvenile opossum, *Didelphis virginiana*

Masked shrew, *Sorex cinereus*

2

Southeastern shrew, *Sorex longirostris*

Smoky shrew, *Sorex fumeus*

Short-tailed shrew, *Blarina brevicauda*

Least shrew, *Cryptotis parva*

Hairy-tailed mole, *Parascalops breweri*

Eastern mole, *Scalopus aquaticus*

Little brown myotis, *Myotis lucifugus*

Southeastern myotis,
Myotis austroriparius

Gray myotis,
Myotis grisescens

Keen's myotis,
Myotis keenii

7

Indiana myotis, *Myotis sodalis*

Small-footed myotis,
Myotis leibii

Silver-haired bat,
*Lasionycteris
noctivagans*

Eastern pipistrelle,
Pipistrellus subflavus

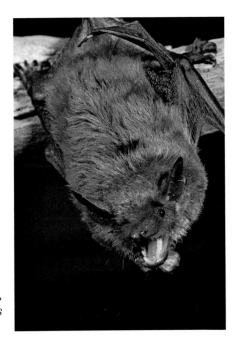

Big brown bat,
Eptesicus fuscus

Red bat, *Lasiurus borealis*, male and female

 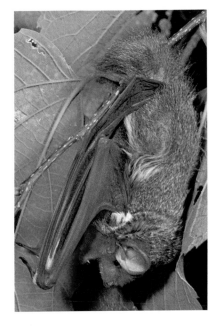

Evening bat, *Nycticeius humeralis*

Hoary bat,
Lasiurus cinereus

Townsend's big-eared bat,
Plecotus townsendii

Rafinesque's
big-eared bat,
Plecotus rafinesquii

Brazilian free-tailed bat,
Tadarida brasiliensis

12

Eastern cottontail, *Sylvilagus floridanus*

New England cottontail, *Sylvilagus transitionalis*

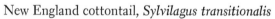

13

KARL MASLOWSKI

Swamp rabbit, *Sylvilagus aquaticus*

Eastern chipmunk, *Tamias striatus*

Woodchuck, *Marmota monax*

Gray squirrel,
Sciurus carolinensis

KARL MASLOWSKI

Fox squirrel, *Sciurus niger*

Southern flying squirrel, *Glaucomys volans*

Beaver, *Castor canadensis*

Marsh rice rat, *Oryzomys palustris*

Eastern harvest mouse, *Reithrodontomys humulis*

18

Prairie deer mouse, *Peromyscus maniculatus bairdii*

Cloudland deer mouse, *Peromyscus maniculatus nubiterrae*

White-footed mouse, *Peromyscus leucopus*

Cotton mouse, *Peromyscus gossypinus*

Golden mouse, *Ochrotomys nuttalli*

Hispid cotton rat, *Sigmodon hispidus*

Eastern woodrat, *Neotoma floridana*

Gapper's red-backed mouse,
Clethrionomys gapperi

22

Meadow vole, *Microtus pennsylvanicus*

Prairie vole, *Microtus ochrogaster*

Pine vole, *Microtus pinetorum*

Muskrat, *Ondatra zibethicus*

Southern bog lemming, *Synaptomys cooperi*

Norway rat,
Rattus norvegicus

House mouse, *Mus musculus*

Meadow jumping mouse,
Zapus hudsonius

Woodland jumping mouse, *Napaeozapus insignis*

Coyote, *Canis latrans*

Red fox, *Vulpes vulpes*

Gray fox, *Urocyon cinereoargenteus*

Black bear, *Ursus americanus*

JOHN MAC GREGOR

28

Raccoon, *Procyon lotor*

Long-tailed weasel, *Mustela frenata*

Mink, *Mustela vison*

Eastern spotted skunk, *Spilogale putorius*

KARL MASLOWSKI

Striped skunk, *Mephitis mephitis*

River otter, *Lontra canadensis*

MARTY STOUFFER

MARTY STOUFFER

Bobcat kittens

Bobcat, *Lynx rufus*

32

Fallow deer, *Cervus dama*

White-tailed deer, *Odocoileus virginianus*

TENNESSEE VALLEY AUTHORITY

ORDER RODENTIA   *Rodents*

The rodents are the most abundant wild mammals in Kentucky, both in number of species and number of individuals. In form and habits they are extremely diverse, but all have chisel-like incisors and no canine teeth. The absence of canines leaves a conspicuous gap in the line of teeth, between the incisors and the molars.

This, the largest mammalian order, is cosmopolitan in distribution. It is represented in Kentucky by 19 genera in 5 families.

4a

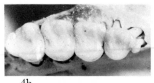 

4b   4b

b. Molar teeth with tubercles in 2 series or else flat-crowned: CRICETIDAE, New World rats and mice, p. 162

FAMILY SCIURIDAE   Squirrels

Members of this family are small to medium-sized, with well-furred tails. The molariform teeth are rooted, and the last 4 are of nearly equal size. Represented in Kentucky by 5 species in 4 genera.

KEY TO GENERA AND SPECIES OF KENTUCKY SCIURIDAE

1. a. Incisors white; tail short and bushy; top of the skull flat; body robust: *Marmota monax*, Woodchuck, p. 139
   b. Incisors yellow; tail elongate, bushy or flat; top of the skull convex; body slender: 2

2. a. With a furred membrane connecting the forelimbs and hind limbs; tail flattened: *Glaucomys volans*, Southern Flying Squirrel, p. 152
   b. No membrane connecting the forelimbs and hind limbs: 3

3. a. Back with prominent stripes: *Tamias striatus*, Eastern Chipmunk, p. 135
   b. Back without prominent stripes: 4

4. a. Total length less than 500 mm; salt-and-pepper gray dorsally; edges and undersurface of the tail whitish: *Sciurus carolinensis*, Gray Squirrel, p. 144

b. Total length more than 500 mm; usually tawny brown, grizzled with gray above; rufous to orange-brown below: *Sciurus niger*, Fox Squirrel, p. 149

Eastern Chipmunk;
Ground Squirrel   PLATE 13

Tamias striatus (Linnaeus)

Recognition: Total length 215–299 mm; tail 78–113 mm; hind foot 32–38 mm; weight 74–129 g. A brown, striped ground squirrel with prominent ears and a flattened, well-haired tail. The pattern on the back is 5 dark stripes and 2 light-buffy stripes. There are prominent internal cheek pouches.

Variation: Kentucky chipmunks are among the darkest known, varying in color from dark russet to chocolate brown. Surprisingly, variation within a population is much greater than geographic variation. Specimens from the mountains of southeastern Kentucky average somewhat darker than those from the rest of the state. Chipmunks from most of Kentucky are referred to the subspecies *T. s. striatus* (Linnaeus). Those from northern Kentucky have been referred to *T. s. ohioensis* Bole and Moulthrop, a race described as being the darkest of this genus. We cannot detect such geographic variation in our collection of Kentucky chipmunks, but we have seen few specimens from that region.

Confusing Species: The chipmunk cannot be confused with any other Kentucky mammal; it is our only striped squirrel.

Kentucky Distribution: Statewide and locally common, this species is unpredictable in its local distribution, being absent from many places where the habitat seems suitable. It is common and sometimes abundant in most Kentucky wood-

Distribution of *Tamias striatus* in the United States

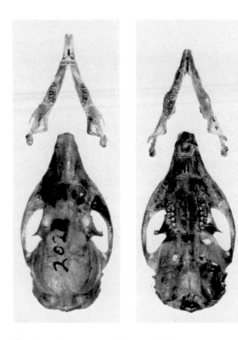

Skull of *Tamias striatus*, x 1.2

lands, as well as in parks, cemeteries, and suburban gardens; the chipmunk is one of our most familiar wild mammals.

Life History: Chipmunks live in burrows, from which they emerge to feed and frolic by day. Their favored habitat is a varied terrain that includes such things as rock walls or cliffs, fallen logs, woodpiles, and hedgerows. They are adequate climbers, frequently going up into trees, but are not as agile as the tree squirrels and, unless hard pressed, seek refuge from danger in their burrows instead of in the trees.

The entrance to a burrow is easily overlooked, because it usually has no dirt piled outside and is often concealed among the leaves. The burrow is several m long (occasionally up to 9 m) and 6 to 9 dm deep. Normally there is more than one entrance. The nest chamber is usually about 3 dm in diameter and is lined with leaves. Larger chambers, used for storing a winter food supply, hold as much as 35 liters of acorns or other food.

Chipmunks eat a wide variety of food, including seeds or fruit of maple, oak, hazel, basswood, hickory, beech, elm, box elder, wild cherry, Virginia creeper, ash, elderberry, and various grasses and weeds. Dried corn, wheat, and oats are also taken, as are blackberries, raspberries, fungi, snails, slugs, insects, and occasional salamanders, frogs, small snakes, and even birds. One was seen to attack, kill, and partially devour a starling in Lee County.

In the fall chipmunks busily store food for the winter. Seeds are forced into the cheek pouches with the forefeet until both cheeks bulge and can hold no more. Then the chipmunk scurries to its burrow to unload the cache in a storage chamber. One such burrow, opened with a spade in a garden in Lexington in spring, contained several liters of sprouting pin oak acorns.

As cold weather approaches, chipmunks spend less time above ground, and by late November most of them are hibernating. Unlike most other hibernating mammals, chipmunks do not get overly fat in autumn; instead, they rely upon their food cache as an energy source, and they awaken every few days to feed. On warm days they venture above ground and are occasionally seen romping in the snow.

By the first warm days in March most Kentucky chipmunks are active above ground. Breeding occurs in March, and a litter of 2 to 7 is born in an underground nest after a gestation period of 1 month. The young are naked and helpless. They grow slowly; their eyes do not open until the young are 30 days old. Ten days later they venture above ground and at 8 weeks of age are on their own.

In Kentucky some chipmunks are born in late summer and are still easily recognized as young when they emerge the following spring. Whether 2 litters are produced or the late litters belong only to those that were too young to breed in their first spring has not been determined.

The home range is small—often less than 0.15 hectare, with most activity taking place within 30 m of the burrow. That they may be familiar with more extensive areas is suggested by the fact that chipmunks are good at returning when displaced as much as 180 m from home.

When feeding, the chipmunk chooses a raised spot, such as a stump, rock, or log, as a lookout point. Piles of shredded acorn shells or other middens betray such sites. If the feeding chipmunk is disturbed, he darts for his burrow with his tail high in the air and emits a series of high-pitched chipping alarm notes, which apparently serve to alert other animals to the danger.

Chipmunks also have a single low-pitched "chuck" sound, uttered as a separate note; it seems to be a communication among contented individuals. An animal will sit quietly in the woods and "chuck" every few seconds for half an hour or more. Often it is answered by others, until the forest may ring with the sound of chipmunks. Because this behavior is most common on pleasant days in the fall, the chucking sound of chipmunks is probably familiar to every upland hunter in Kentucky, but surprisingly few of them know the origin of the sound.

Chipmunks are the prey of most of our larger predators

that are active by day. In the city and suburbs house cats are probably their worst enemy. Elsewhere they are also fed upon heavily by hawks, weasels, and rat snakes. Many are struck down by automobiles, and dogs kill numbers of them.

Chipmunks are of little economic significance, but because of their attractiveness they are generally considered desirable. Occasionally an individual learns to dig up corn that has been freshly planted, and this has given rise to the saying of the old-time farmer planting 5 kernels of corn to the hill: "One for the chipmunk, one for the crow, one for the cutworm, and two to grow." In some years chipmunks become abundant enough to be a nuisance in the garden. In a Lexington garden where chipmunks were digging up corn in the summer of 1973 we removed 13 adults, although never more than a single chipmunk was seen at one time.

Woodchuck; Groundhog   Plate 14

Marmota monax (Linneaus)

Recognition: Total length 418–665 mm; tail 100–155 mm; hind foot 68–88 mm; weight 2.7–5.4 kg (heaviest in autumn). This is a large, stout, brownish animal with a short snout and a short, furry tail. Color varies from nearly black to grizzled gray to dark brown; the underparts are lighter, often with a reddish wash, and the feet and legs are dark brown to nearly black. The toes have well-developed claws.

Variation: No geographic variation is evident in Kentucky. Our subspecies is *M. m. monax* (Linnaeus).

Confusing Species: The adult woodchuck is too large to be confused with any of its squirrel relatives. A young woodchuck is readily distinguished from the other squirrels by its more massive body and its short tail.

Distribution of *Marmota monax* in the contiguous United States

Skull of *Marmota monax*, x 0.4

Kentucky Distribution: Statewide, but more numerous in the eastern half of Kentucky. In the Purchase the woodchuck is scarce and local. The woodchuck is not nearly so common in Kentucky as in the northeastern states.

Life History: The woodchuck is an animal of the forest edge, fencerows, and roadsides; although present in the deep forest, it is much less common there. It is most active in early morning and late afternoon, averaging only about 3 hours a day above ground. Although generally diurnal, it has been reported occasionally to feed at night.

The home is a burrow with a fresh mound of earth at the entrance. The entrance hole usually measures about 150–180 mm high and 180–200 mm wide. The woodchuck is an excellent digger, and the burrows are often extensive, reaching a depth of 1.5 m and covering a distance of 9 m. Usually there are 2 or 3 entrances, and often one of these escape holes is well concealed and has no mound of earth beside it.

One or more branches of the tunnel lead to a chamber containing a bulky nest of dried grass. Here the 4 or 5 young (litter sizes of 8 and 9 have been reported) are born in April after a gestation period of 3 or 4 weeks. The young are blind and remain in the den until their eyes open, at about 4 weeks of age. From a birth weight of about 30 g they grow to 900 g in about 8 weeks.

Woodchuck burrow.
Steel fence post
suggests size.

Woodchucks are almost strictly vegetarians. Less than 1% of the food is animal matter, consisting of such insects as grasshoppers and June beetles; however, one was reported to have eaten a small nestling bird. The favored food includes a variety of weeds, such as dandelion and plantain. Woodchucks are especially fond of legumes, such as clover, alfalfa, vetch, peas, and beans. They also eat buckwheat, oats, wheat, corn, beet and turnip tops, cabbage, kale, cantaloupe, raspberries, strawberries, cherries, and apples. Although one living next to your garden can be a nuisance, woodchucks seldom cause trouble in Kentucky.

The woodchuck is a game animal, large enough to provide a substantial amount of meat. When the flesh is soaked in salt water overnight and then boiled and fried, it is delicious. We are unable to substantiate the common tale that the musk glands, under the forelegs, must be removed as soon as possible after the animal is killed.

Groundhogs are also valuable in providing denning sites for other game animals and furbearing animals, such as rabbits, skunks, and opossums. Rabbit populations have been shown to be affected by the number of burrows available.

In late summer woodchucks get extremely fat in preparation for hibernating. As fall advances they spend less and less time above ground, and by the end of October nearly all are in hibernation.

The hibernation chamber is a dead-end nest chamber—sealed off with dirt so that the woodchuck is not disturbed by other animals, which use the rest of the burrow system in winter. The body temperature drops, the heart rate slows to as few as 4 beats per minute, and the animal sleeps through the winter. With the first warm days in February a few woodchucks appear above ground to feed on dandelion leaves and whatever other greenery is available.

Contrary to legend, a woodchuck's shadow seen on February 2 has no bearing on the weather of the next six weeks.

Like many other members of the family Sciuridae, the woodchuck is a competent climber. When chased by men or dogs it does not hesitate to climb a tree. Occasionally one will climb to feed on pawpaws or leaves or simply to lie on a horizontal limb and bask in the sun.

When a woodchuck is taken by surprise it sometimes emits a very loud, shrill whistle and dives for its burrow. This characteristic sound has given rise to the name "whistle pig."

Remarks: Baby groundhogs are commonly captured. They tame easily, present few problems in care and feeding, and generally make good pets.

Woodchucks are good swimmers when, occasionally, they take to the water. One was observed to swim across a river about a kilometer wide.

The front teeth of rodents continue to grow throughout life, compensating for wear due to gnawing. In woodchucks it is not at all uncommon for these teeth to be so damaged as to fit improperly, resulting in one or more of them not wearing normally. The resulting unimpeded growth prevents the animal from eating, or even pierces the brain. Malformed incisors occur in more than 1% of woodchucks.

We can understand the hunting of groundhogs with small-caliber, high-velocity, 'scope-sighted rifles, provided it is done legally. We cannot understand, however, the urge that drives so many hunters to hang the carcass on a roadside fence. It is a useless, wasteful end of a most interesting mammal.

The woodchuck ranks third in the number of game mammals harvested in Kentucky, and it also ranks third as the objective of hunting trips. The Kentucky Department of Fish and Wildlife Resources estimates that an average of 267,500 woodchucks were taken annually by hunters in the 1964–65 through 1970–71 seasons.

Malocclusion in
a woodchuck skull

Gray Squirrel   PLATE 14

Sciurus carolinensis Gmelin

Recognition: Total length 430–500 mm; tail 210–240 mm; hind foot 60–70 mm; weight 400–710 g. A large, gray tree squirrel with a long, flattened, bushy tail. Color above varies from gray to brownish-gray; the underparts usually are white, but this is variable, the occasional extreme being dark brown. Melanism (black) and albinism (white) seem more common in the gray squirrel than in other Kentucky mammals.

Variation: Little geographic variation is evident in Kentucky. Our subspecies is *S. c. carolinensis* Gmelin. In winter the squirrels of the Cumberland Plateau and the Bluegrass develop a white spot on the back of the ears like the more northern *S. c. pennsylvanicus* Ord.

Confusing Species: The fox squirrel, *S. niger*, is the only other similar tree squirrel in Kentucky. The 2 species usually are rather easy to distinguish, because the larger fox squirrel has yellowish to orange underparts and feet and a scattering of yellow hairs in the tail.

Kentucky Distribution: Statewide; common. This is the common squirrel of forested regions throughout the state, as well as of city parks and suburban yards.

Life History: The oak–hickory forests of Kentucky are among the favored haunts of the gray squirrel. The size of the population usually is proportionate to the crop of acorns and nuts upon which *S. carolinensis* feeds. In a young forest, where hollow trees are absent, the number of squirrels can be increased by providing nesting boxes.

Squirrels are strictly diurnal. They can be seen at any time of day but are most active in the early morning and from late afternoon to early evening. In the fall, when the hunter is

Distribution of *Sciurus carolinensis* in the United States

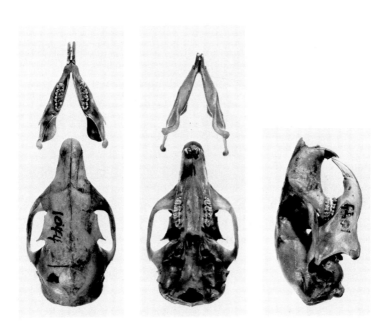

Skull of *Sciurus carolinensis*, x 0.8

Tree nest of a gray squirrel

seeking them, they are often quiet and shy. Signs of their presence are cuttings of acorn shells or nut husks beneath a tree in which they have been feeding. Squirrel cuttings can be distinguished from those of chipmunks, which tend to make piles on stumps or rocks.

A squirrel that has been disturbed will sometimes utter a harsh warning call, which hunters call a "bark." Squirrels that are not hunted or otherwise threatened are often much more vociferous than those subjected to hunting. With tail vibrating and with a strong flip at each note, the squirrel utters a guttural "kwa-ak, kwa-ak, kwa-ak, kwa-ak, ak-ak-ak-ak-ak-ak." Sometimes another squirrel responds, and the two may call many times.

Although the gray squirrel is most at home in the trees, it commonly forages on the ground for nuts or other food or romps there in play or in the mating chase. A new source of food, such as anything put out for birds, is quickly found and exploited. Sometimes squirrels will discover corn in the garden or peaches on a fruit-laden tree and make a nuisance of themselves.

Squirrels are active throughout the year. Tracks in the snow leading from one tree to another indicate that they have little fear of the cold. Rainy or windy days are more likely to keep them inactive.

For its winter home a squirrel chooses a hollow tree, where it builds a cozy nest of shredded bark and other plant fibers. Often the same nest is used to rear the young. In summer squirrels build large, bulky nests of leaves, high above the ground in the branches of a tree. Such nests are used especially by young squirrels.

In Kentucky the breeding activities of squirrels are highly seasonal. From August to November the testes of the males are tiny and are located within the abdominal cavity; close inspection is required to recognize the animals as males. In late fall the testes begin to enlarge and descend into the scrotum, and in January the peak of reproductive activity is reached. As the males become reproductively active they begin to chase the females—a ceremony that lasts 1 to 3 days before mating.

Gestation takes 43 to 45 days, and the young, usually numbering 3 or 4, are born in March or April. They weigh about 14 g at birth and are naked and blind. Their eyes open after about 32 days, and the young begin to appear outside the nest chamber at 45 days of age. They are weaned before they are 2 months old.

Reproductive capacity in the males, after reaching another low point (degeneration of the testes and their migration into the abdominal cavity) in April, begins to develop anew in June. By July the second breeding season is in full swing, and a second litter is produced in September.

Although nuts and acorns are its major food items, the gray squirrel enjoys a varied diet. Seeds and fruits of maple, elm, hornbeam, arrowwood, dogwood, hackberry, sassafras, beech, wild cherry, mulberry, and viburnum, among others, are eaten. Buds of maple, elm, and oak are favored. The inner bark of maple and elm is sometimes eaten in winter. Nearly anything put out for the birds in winter is relished by squirrels, but they are especially fond of sunflower seeds. They also take some animal matter, including insects and, occasionally, eggs and young birds. Squirrels frequently gnaw bones, from which they apparently obtain needed minerals.

When squirrel populations are high, mass migrations involving thousands of animals occur. The pattern is for exceptional numbers of squirrels to survive a winter following a good crop of acorns and nuts. Then, if a poor mast crop develops the next summer, the squirrels suffer from hunger and begin to move. Before the turn of the century, when the native chestnut was a bountiful food item, squirrels were much more abundant, and spectacular migrations were reported. The most recent squirrel migration in Kentucky was in 1968. While the squirrels were moving, any drive along country highways would reveal that many had been killed by traffic.

Most migrating squirrels fall victim to accidents or predators. Food is no more abundant elsewhere than at home, yet they strike out in one direction and travel until they meet their fate. Even water does not stop them. In quiet water a squirrel can swim well for about 3 km, but if the water surface is rough it soon drowns.

Squirrels are the most important game animals in Kentucky. The Kentucky Department of Fish and Wildlife Resources estimates that an average of 1,309,000 were taken annually in the 1964–65 through 1970–71 hunting seasons. Gray squirrels and fox squirrels were not separated in the estimate.

Sketch of squirrel tracks in snow, x 0.3: left, S. *carolinensis*; right, S. *niger*

Fox Squirrel PLATE 15

Sciurus niger Linnaeus

Recognition: Total length 510–565 mm; tail 218–265 mm; hind foot 70–78 mm; weight 675–1,000 g. This is the largest of our tree squirrels. Upperparts grizzled gray, often with a wash of yellow or orange; underparts pale yellow to bright orange. We have one Kentucky specimen with black underparts.

Variation: No geographic variation is evident in Kentucky. Our subspecies is *S. n. rufiventer* E Geoffroy St.-Hilaire.

Confusing Species: Our only similar tree squirrel is the more common gray squirrel, *S. carolinensis.* Color patterns differ, in that the fox squirrel has yellow to orange underparts and the gray squirrel is white to brown below. The tail of the gray squirrel contains many white hairs; in the fox squirrel such hairs are yellow. The tiny premolar usually present in front of the first large upper cheek tooth in the gray squirrel is missing in the fox squirrel.

Kentucky Distribution: Statewide; fairly common. Most frequent in the Bluegrass and on the Mississippian Plateau; scarce and local elsewhere.

Life History: The fox squirrel is an animal of open country in Kentucky. Its favored habitat is the scattered oak, hickory, and walnut trees along the fencerows of pastures and cultivated land.

This is a strictly diurnal species, most active in the morning and early afternoon. It spends more time on the ground than the gray squirrel and often forages far from the nearest tree. The home range is generally a hectare or 2, but at times of food scarcity individuals may range over 8 hectares. When frightened a fox squirrel will run to the nearest tree, at a speed of up to 29 km per hour.

Distribution of *Sciurus niger* in the United States

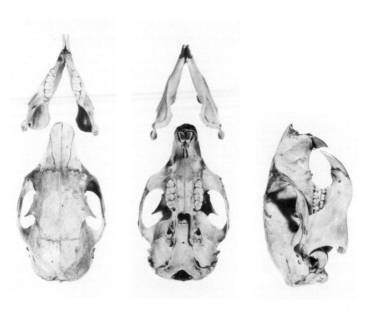

Skull of *Sciurus niger*, x 0.7

Typical fox squirrel habitat in Kentucky

This is generally a quieter squirrel than the gray, so the hunter must rely more on his eyes than his ears when seeking it. The occasional barking and scolding notes are similar to those of the gray squirrel but can be distinguished by someone who is quite familiar with both: the fox squirrel's bark generally contains fewer notes.

Acorns, hickory nuts, and walnuts are the favored food. Various fruits and seeds are also eaten. Cornfields may be raided and the ears carried off into the woods. Mushrooms, an occasional insect, and various other edibles round out the diet. Like other squirrels, S. niger stores nuts for winter use, burying them in the ground and beneath the leaves.

The home of the fox squirrel is a leafy nest, usually about 40 feet above the ground in the crotch of a tree. Hollow trees are also favored, especially for raising the young and as shelters in the coldest winters. A fox squirrel often will use more than one nest.

Mating occurs in January and February and again in May and June. The gestation period is 44 days. Litter size is usually 3 to 5, but as many as 7 fetuses have been found in a fox squirrel. The young are naked at birth, and their ears and eyes are closed. The ears open at about 25 days of age; the eyes, at about 32 days. At 2 months of age the young squirrels venture from the nest, and they are weaned at about 3 months. A fox squirrel born in spring is ready to breed the following spring.

Remarks: If captured when young, a fox squirrel makes an excellent pet. It is gentler than the gray squirrel and as interesting in captivity as a chipmunk or a groundhog.

Fox squirrels have few natural enemies. Young ones are occasionally taken by some of our larger predatory birds and mammals, and dogs and cats kill a few. Far more fox squirrels fall to automobiles and guns. Being larger and more attractive than the gray squirrel, the fox squirrel is prized by the hunter.

Southern Flying Squirrel   PLATE 15
Glaucomys volans (Linnaeus)

Recognition: Total length 211–253 mm; tail 81–120 mm; hind foot 28–33 mm; weight 70–98 g. This is a beautiful little squirrel with large eyes and silky fur. A loose fold of skin extends from the wrist of the forefoot to the ankle of the hind foot on each side; this is the gliding membrane. Color is gray to brown above and white below; the 2 colors are separated by a black line along the edge of the gliding membrane.

Variation: Two subspecies are recognized in Kentucky. *G. v. volans* (Linnaeus), occurs throughout most of the state. The darker southern race, *G. v. saturatus* Howell, has been reported from Big Black Mountain, in Harlan County, and probably occurs across the southern part of the state.

Confusing Species: The flying squirrel cannot be confused with any other Kentucky mammal. It is similar to the northern flying squirrel (*G. sabrinus*), which has not been found in Kentucky but which may occur on the highest ridges in the south-

Flying squirrels often
inhabit woodpecker holes.

eastern part of the state. The northern species is larger and
can be recognized by the dark base of the hairs on the belly.
In *G. volans* the hair is white to the base.

Kentucky Distribution: Statewide. Most common in the
southern and western counties.

Life History: Because it is strictly nocturnal, the common
little flying squirrel is seldom seen by man. Flying squirrels
spend the day in a nest in a woodpecker hole or other tree
cavity, often near a stream or pond. They emerge at dusk and
forage during the night, feeding upon nuts, seeds, and insects.

Upon emerging from the nest site in the evening, a flying
squirrel launches itself into a glide, which carries it to a lower
spot on a neighboring tree. It then ascends the tree, explores
for food, and soon glides off to another. Thus, the "flight"

Distribution of *Glaucomys volans* in the United States

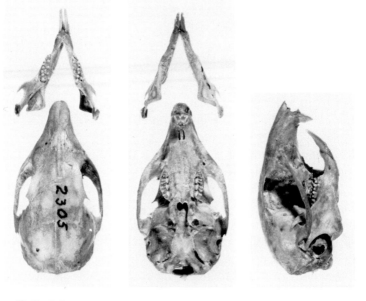

Skull of *Glaucomys volans*, x 1.3

of a flying squirrel is quite different from that of a bat or bird. Flying squirrels are most easily located by listening for their calls at night. When foraging they frequently emit weak twittering and chirping—bird-like—sounds, which are seldom noticed unless one is listening for them. One can hear them in nearly any forest or woodlot in Kentucky. Although flying squirrels occasionally forage on the ground for short periods, they are much more likely to be in a tree.

Mating occurs in January or February. After a gestation period of 40 days, the young are born in a nest cavity in a tree. The nest consists of leaves and shredded bark, with a lining of finely shredded inner bark.

Three or 4 young comprise the usual litter, but as many as 6 are sometimes produced. At birth the young are blind and hairless, weigh only 2 or 3 g, and are about 50 to 75 mm in length. They grow rapidly, doubling their weight within a week. Their eyes open at 4 weeks of age, and the young begin foraging for themselves at 6 weeks. A second litter is born in August.

Food consists of hickory nuts, walnuts, beechnuts, and acorns; seeds of wild cherry, hackberry, and maple; fruits of grape, apple, and viburnum; and a variety of insects. The eggs and young of birds are sometimes taken. We have captured flying squirrels in rat traps nailed to trees and baited with bacon or other meat. Like the other squirrels, this species stores food in tree cavities.

Flying squirrels are active throughout the winter. At this season they become quite social; many individuals may sleep in the same nest. As many as 22 have been found in a single hollow tree in Wisconsin, but half a dozen or so is a more usual find. If you pound on a tree occupied by flying squirrels, they will awaken and look out. Continued or severe disturbance may cause them to glide off to other trees.

Remarks: Like other members of the family Sciuridae, this species makes a good pet. Flying squirrels are very attractive and gentle; if handled properly they seldom bite.

The flying squirrel is of little economic significance. It is not hunted as game, and it rarely damages crops. Occasionally it nests in the attic of a house and becomes a noisy nuisance. Cats and owls seem to be the major predators on flying squirrels. Rat snakes probably also capture them.

Family Castoridae Beavers

Large, stout, heavy-bodied aquatic rodents with a broad, flat, scaly tail and webbed hind feet. The molars are rootless. Represented in Kentucky by a single species.

Beaver   Plate 16
Castor canadensis Kuhl

Recognition: Total length 900–1,170 mm; tail 300–400 mm; hind foot 165–185 mm; weight 11–32 kg. A large, heavy-bodied, dark-brown, aquatic rodent with webbed hind feet, a broad, flat tail, and dense, silky fur.

Variation: No geographic variation is recognized in Kentucky. Our native subspecies is *C. c. carolinensis* Rhoads.

Confusing Species: The beaver is sometimes confused with the much smaller muskrat, with which it shares its aquatic habitat. A beaver is easily recognized by its flat tail, which is about 150 mm broad.

Kentucky Distribution: Statewide; fairly common.

Life History: Our largest and perhaps most interesting rodent, and one of our largest furbearers, the beaver has had a profound

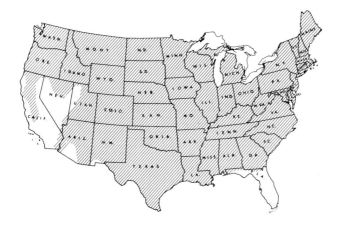

Distribution of *Castor canadensis* in the contiguous United States

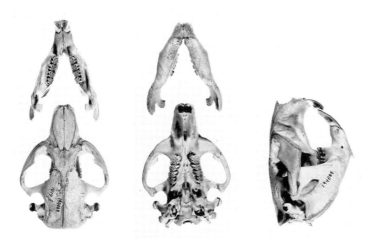

Skull of *Castor canadensis*, x 0.2

Beaver dam in Powell County

influence on the history of North America. Abundant and prolific, with fur of the highest quality, the beaver was intensively sought by the early white traders and settlers. Establishment of the Hudson's Bay Company, in Canada, and the early explorations of the Canadian Northwest and the Rocky Mountains resulted primarily from the demand for beaver fur.

So intensively were beavers sought that they were nearly exterminated in the last decades of the 19th century. Protective legislation enacted by the states early in the present century enabled the species to survive in some areas. As numbers increased, beavers were shipped from Wisconsin to states where they had disappeared. Kentucky was one of these states. As recently as 25 years ago beavers were still scarce and local in Kentucky (and the other eastern states); today the species is probably nearly as widespread and common as it was before the arrival of the white man.

In our rivers and impoundments beavers give little evidence of their presence; the home is a burrow in the bank, entered from beneath the water and hidden from sight. When a beaver takes up residence in a small stream, however, the first thing he does is build a dam and a lodge. The dam creates a pond, which serves to protect the animals from their enemies and which can be used to float sticks that have been cut for storage as winter food.

In building a dam a beaver piles sticks across a channel and then shoves mud between them from above. The process is continued until there is a long ridge of sticks with a layer of mud sloping away from them on the upper side. The pond thus created is quite variable in size but sometimes covers a hectare or more. Some beavers build several dams along the course of a small stream.

The dam sometimes provides good fishing for man. Chubs and sunfish often live in beaver ponds, and in the northern states they are favored by brook trout. On the other hand, beaver dams can sometimes be a nuisance, flooding fields and roads.

If a dam is breached, the beavers quickly repair it. A remarkable amount of repair work can be done in a single night by a pair of these nocturnal animals.

In a deep part of the pond (usually near the middle) the beavers build their lodge. This consists of a pile of sticks and mud, the top of which is well above the water. The part above the water contains a single room, 6 to 10 dm high and 10 to 15 dm wide; there the animals live. Entrances are from beneath the water. As many as 8 to 10 beavers may live in a single lodge. When the pond is frozen over and the part of the

Beaver lodge

Tree felled by beavers

lodge projecting through the ice is frozen, the beavers are safe from any predator save man.

The only vocal sound adult beavers make is a low grunting when annoyed. Young beavers, however, are rather vociferous, producing a variety of cries, whines, and playful sounds.

Although beavers are nocturnal, they are occasionally seen swimming by day, especially in the evening or early morning. If startled the beaver will dive, at the same time giving the danger signal: a slap on the water surface with its flat tail. This produces a remarkably loud sound, which sometimes can be heard for a kilometer.

One of the best-known habits of beavers is tree-cutting—made possible by their large incisor teeth and powerful jaw muscles. A beaver at work produces such a gnawing and scraping sound that it can sometimes be heard at a distance of 100 meters or more. Small trees are preferred, but it is not uncommon for beavers to fell one that is 30 cm in diameter; and there is record of a willow 84 cm in diameter being downed by beavers in Wisconsin. Beavers commonly attack a large tree but do not finish cutting it. Most trees cut by beavers are near the stream where they live; they seldom travel more than 100 meters to harvest trees.

Trees that are felled are cut into manageable pieces and used to build dams or lodges or as food. In summer beavers eat sedges, rushes, and other water plants, as well as bark, leaves, and twigs. In winter they feed on bark from branches they have stored underwater near the lodge or bank den.

Beavers are reported to be monogamous and to remain paired for life. Mating takes place in January or February, and the gestation period is 4 months. The single litter per year consists of 1 to 8 kits; usually there are 4 or 5. Young animals do not breed until their second year.

Beavers are born with the body furred and eyes open; they are about 3 dm long and weigh 450 g. They grow rather slowly and weigh only 2.5 to 3.5 kg when 3 months old. At the end of a year the youngsters weigh 12 to 14 kg. The parents and young remain as a family unit until the young are sexually mature, at 2 years of age.

Muskrats commonly associate with beavers and are sometimes mistaken for young beavers. They sometimes even live in the beaver lodge and eat scraps left by the beavers.

The beaver has few natural enemies. Its large size and massive teeth make it a formidable adversary of predators. A beaver can easily whip a dog in the water.

Remarks: Man is the only significant threat to beavers. Valued for their fur, beavers are trapped extensively. Populations are now well managed, and game laws prevent a recurrence of the near-extinction from trapping.

Beavers are generally considered beneficial. Their dams aid in soil and water conservation, are attractive to wildlife, and are pleasing to the eye. Sometimes dams are placed where they are a nuisance, and the beavers must be removed. They can also be a problem where corn is grown near a colony, for the animals are fond of this crop and can be very destructive.

In the 5 trapping seasons 1968–69 through 1972–73 fur-buyers in Kentucky purchased 690 beaver pelts—averaging 138 per year but varying widely, from 46 in 1970–71 to 289 in 1972–73. Average price per pelt was $4.68, but the price ranged from $3.17 in 1970–71 to $6.07 in 1971–72.

FAMILY CRICETIDAE   New World Rats and Mice

An assemblage of various-sized but generally small rodents with never more than 12 molars—3 above and 3 below on each side—and no premolars. The molars are either flat-crowned or tuberculate, but if tuberculate the tubercles are always in 2 rows. Although essentially adapted to a vegetable diet, many cricetids are omnivorous. Represented in Kentucky by 14 species in 10 genera. Some authors include this family with the Old World family Muridae.

KEY TO GENERA AND SPECIES OF KENTUCKY CRICETIDAE

1. a. Molars with 2 longitudinal rows of tubercles or, if flat-crowned, with prisms on grinding surfaces not arranged in alternating triangles; tail long to moderate, at least ⅓ of total length (Subfamily Cricetinae): 2

   b. Molars flat-crowned, with prisms on grinding surfaces arranged in alternating triangles; tail usually short, less than ⅓ of total length except in *Ondatra* (Subfamily Microtinae): 10

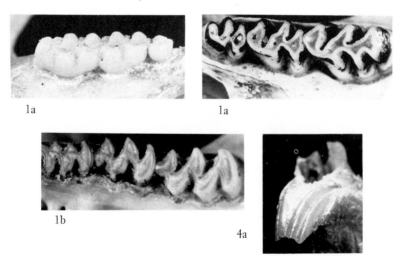

1a   1a

1b

4a

2. a. Molars flat-crowned; size large, total length more than 225 mm: 3

   b. Molars tuberculate; size usually smaller, total length usually less than 225 mm (except *Oryzomys*, whose length may reach 305 mm): 4

3. a. Total length less than 300 mm; tail moderate, 60-90% of body length; hind foot 28–34 mm: *Sigmodon hispidus*, Hispid Cotton Rat, p. 189

   b. Total length more than 300 mm; tail longer, 75–95% of body length; hind foot 35–46 mm: *Neotoma floridana*, Eastern Woodrat, p. 192

4. a. Incisors grooved: *Reithrodontomys humulis*, Eastern Harvest Mouse, p. 168

   b. Incisors not grooved: 5

5. a. Tail usually more than ½ total length, with a tiny tuft of hair at the tip; generally with a dark dorsal stripe and white-edged ears. Occurs only in the southeastern mountains: *Peromyscus maniculatus nubiterrae*, Cloudland Deer Mouse, p. 175

   b. Tail usually less than ½ the total length; usually without a dark dorsal stripe and white-edged ears (except in *Peromyscus maniculatus bairdii*): 6

6. a. Back and sides golden in color; no distinct line of separation between side and belly fur: *Ochrotomys nuttalli*, Golden Mouse, p. 185

   b. Back and sides not golden, although sometimes fawn-colored; a distinct line of separation between side and belly fur: 7

7. a. Tail distinctly bicolor, with a tiny tuft of hair at the tip; ears white-edged; total length about 150 mm; tail 60 mm or less; hind foot about 18 mm: *Peromyscus maniculatus bairdii*, Prairie Deer Mouse, p. 171

b. Tail not distinctly bicolor; ears not white-edged; size larger: 8

8. a. Size smaller, usually less than 190 mm in total length; hind foot usually less than 22 mm: *Peromyscus leucopus*, White-footed Mouse, p. 177

   b. Size larger, total length usually more than 190 mm: 9

9. a. Size smaller, total length less than 210 mm; tail less than 100 mm; hind foot usually more than 22 mm: *Peromyscus gossypinus*, Cotton Mouse, p. 182

   b. Size larger, total length more than 220 mm; tail more than 100 mm: *Oryzomys palustris*, Marsh Rice Rat, p. 165

10. a. Tail nearly ½ the total length and laterally compressed; size larger, total length more than 250 mm: *Ondatra zibethicus*, Muskrat, p. 212

    b. Tail short, less than ⅓ the total length, terete; size smaller, total length less than 250 mm: 11

11. a. Incisors with a shallow groove; tail about as long as the hind foot: *Synaptomys cooperi*, Southern Bog Lemming, p. 217

11a

14a

14b

b. Incisors not grooved; tail usually longer than the hind foot: 12

12. a. Color reddish above: 13
    b. Color not reddish above: 14

13. a. Fur coarse; tail usually more than 30 mm: *Clethri-onomys gapperi*, Gapper's Red-backed Mouse, p. 196
    b. Fur soft; tail usually less than 30 mm: *Microtus pine-torum*, Pine Vole, p. 208

14. a. Fur above grizzled gray; tail 24–41 mm; 5 plantar tubercles; crown of middle upper molar with 4 irregular loops: *Microtus ochrogaster*, Prairie Vole, p. 204
    b. Fur above dull brown; tail 32–64 mm; 6 plantar tubercles; middle upper molar with 4 triangles and a posterior loop: *Microtus pennsylvanicus*, Meadow Vole, p. 199

Marsh Rice Rat   PLATE 17

Oryzomys palustris (Harlan)

Recognition: Total length 226–305 mm; tail 108–156 mm; hind foot 28–37 mm. A small, slender rat with a long tail. Brown to grayish-brown above and nearly white below.

Variation: No geographic variation is recognized in Kentucky. The subspecies in this state is O. p. palustris (Harlan).

Confusing Species: Similar to the common Norway rat, *Rattus norvegicus*, but much smaller; easily confused with a young Norway rat. The young Norway rat is of heavier build; has larger hind feet; and has a stouter tail, which is nearly uniform in color, whereas the slender tail of the rice rat is bicolored, being brown to gray above and whitish below. The molars of the rice rat have only 2 rows of tubercles.

The rice rat is easily distinguished from the much larger woodrat, which has a furry tail. Other similar species include the voles (*Microtus*), which have short tails, and the cotton rat, *Sigmodon hispidus*, which is heavier and darker and has a shorter tail.

Kentucky Distribution: The western part of the state, eastward to Trigg County. Occurs throughout the Purchase and around Kentucky and Barkley reservoirs. Also known from Barbourville, in southeastern Kentucky, on the basis of 3 specimens taken before 1918.

Life History: The rice rat is a resident of lowland marshes. In the counties west of the Tennessee River, where they are fairly common, they can be found beside almost any roadside ditch or stream where grasses and shrubs provide sufficient cover.

Rice rats do not make runways as distinct as those of meadow mice and cotton rats. In our experience, in Kentucky they use poorly defined runways and sometimes have none at all; in other states we have found them using cotton rat runs.

Rice rats are at home in the water, where they swim and dive with ease; they are difficult to capture in the water. In the salt marshes of the coastal states rice rats live in spherical nests made of shredded grass and placed about 3 dm or so above high-water level in the vegetation of the intertidal zone; when frightened the rats jump into the water and dive to safety. In inland regions the nests are often in shallow burrows on higher but soft ground a meter or so back from the water.

The breeding season of rice rats in Kentucky is unknown, but in the southern coastal states it extends from February to November. One to 6 young are born, after a gestation period of 25 days. They are blind and naked at birth but grow rapidly. Their eyes open on the sixth day, and the young are weaned at 11 days. They are sexually mature within 2 months.

Rice rats are often carnivorous, feeding upon almost any type of animal material available. Insects, birds and their eggs,

Distribution of *Oryzomys palustris* in the United States

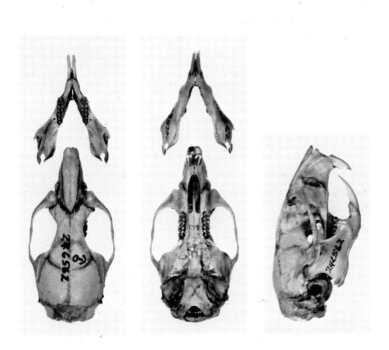

Skull of *Oryzomys palustris*, x 1.6

Locality records of *Oryzomys palustris* in Kentucky

crayfish, young turtles, and other animals are eaten. If animal material is not readily available, rice rats eat the seeds of a variety of plants. Their common name comes from their habit of feeding on rice in newly planted fields. They also feed on the rice crop and on grains scattered during harvest.

Rice rats are nocturnal and seldom seen by man. They are active throughout the year; tracks in the snow reveal their complex and extensive foraging pattern.

Hawks and owls, mink, weasels, and foxes prey upon rice rats. The cottonmouth moccasin is said to prey heavily upon them.

In Kentucky, a state where rice is not raised, the rice rat is of little economic importance. It is a southern species, which barely ranges into the state and is simply an interesting minor member of our fauna.

Eastern Harvest Mouse   PLATE 17

Reithrodontomys humulis (Audubon and Bachman)

Recognition: Total length 107–150 mm; tail 45–68 mm; hind foot 15–17 mm; weight about 8 g. A little brown mouse with small ears and small, beady eyes. Upperparts dark brown; underparts gray. The upper incisors are grooved.

Variation: No geographic variation is evident in Kentucky. Our subspecies is *R. h. humulis* (Audubon and Bachman).

Confusing Species: The common house mouse, *Mus musculus*, is gray rather than brown and does not have grooved upper incisors. Its fur has a greasy texture, and it is larger than our harvest mouse. The white-footed mice, *Peromyscus leucopus* and *P. maniculatus*, also lack grooved incisors, and they have white underparts.

The western harvest mouse, *R. megalotis*, has extended its range eastward into Indiana and may someday appear in Kentucky; it is now known from Leachville, Arkansas, less than 105 km from our border. It is larger than our harvest mouse and is gray in color.

Kentucky Distribution: Statewide; uncommon. Local in weed-fields.

Life History: The little harvest mouse is found in fields of tall, dense weeds. Although it is rarely as common as white-footed mice, persistent trapping in any such field in Kentucky will probably yield a few.

Harvest mice are nocturnal animals, rarely seen by man. They spend much of their time foraging for weed seeds. The soft, spherical nest is made of shredded grass and other plant fibers and is hidden in the weeds. It is usually made on the ground under the shelter of a stone or clod, but sometimes it is supported by weeds, several centimeters above the ground.

The adults are solitary or go in pairs in the warm season. In winter they commonly nest in groups during the daylight hours. Nest boxes or cans filled with cotton and placed in a weed-field may be used by the mice; 4 to 6 are sometimes found in the same nest.

In a study at Oak Ridge, Tennessee, harvest mice were found to breed from late spring to late fall, with a decrease in breeding activity in midsummer. Litter size was usually 3 or 4, but one female gave birth to a litter of 8.

Distribution of *Reithrodontomys humilis*

Skull of *Reithrodontomys humulis*, x 2.5

As with other small mammals, the life-span in the wild is short. Most individuals survive less than a year; but the large number of offspring maintains the species. The home range is quite variable among individuals. Some stray only a few dozen meters from the nest; others commonly travel 100 meters or more.

Prairie Deer Mouse   PLATE 18

Peromyscus maniculatus bairdii (Hoy and Kennicott)

Recognition: Total length 116–154 mm; tail 40–58 mm; hind foot 17–20 mm; weight 16–26 g. A small deer mouse with a relatively short, bicolored tail, white-edged ears, and small hind feet. Upperparts brown to gray (juveniles always gray); underparts white. The tail is sharply bicolored, being gray above and white below.

Variation: There are two subspecies of *P. maniculatus* in Kentucky. Because they are quite different from one another and are separated geographically, we are treating them in separate accounts.

Confusing Species: The prairie deer mouse is most similar to the white-footed mouse, *P. leucopus;* occasional individuals are difficult to identify. Most adults can be separated on the following basis: in the prairie deer mouse the hind foot is usually less than 19 mm long, the tail less than 60 mm long, the total length less than 155 mm, and the ear (from the notch) less than 17 mm long. Adult *P. leucopus* nearly always exceed these measurements.

The prairie deer mouse is also similar to the harvest mouse, *Reithrodontomys humulis,* with which it is often found in Kentucky. Deer mice are white rather than gray beneath and do not have grooved upper incisors.

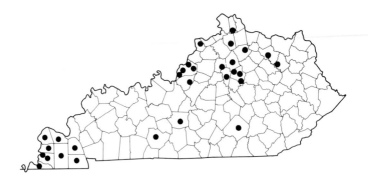

Locality records of *Peromyscus maniculatus bairdii* in Kentucky

Kentucky Distribution: Open weedfields, grasslands, and other agricultural land throughout most of the state. Common in the western and central regions, becoming local in the valleys and weedy hillsides of the Cumberland Plateau. Apparently absent from the southeastern mountains and adjacent parts of the Cumberland Plateau.

Life History: Any area that has been cleared of trees and allowed to grow up in weeds or grass is a good place to look for the prairie deer mouse. It is sometimes the most abundant mammal in weedfields, roadbanks, fencerows, and abandoned agricultural land. Even cultivated land sometimes harbors a population of these mice. We have found them in alfalfa fields and in weedy cornfields after the harvest. A general preference is for dry uplands.

Prairie deer mice are nocturnal and seldom seen by man. Even in favorable habitats, where populations may be as high as 25 per hectare, they leave little sign of their activities. A small and inconspicuous hole, often leading into a mole tunnel, gives access to a nest slightly below the ground surface; there the mouse spends the day. If cover, such as boards, branches, or trash, is available, the nest may be placed beneath it and thus easily found. In warm weather these mice often inhabit nests on the surface, hidden away in a clump of grass or weeds.

▨ *P. m. bairdii*

Distribution of *Peromyscus maniculatus* in the contiguous United States

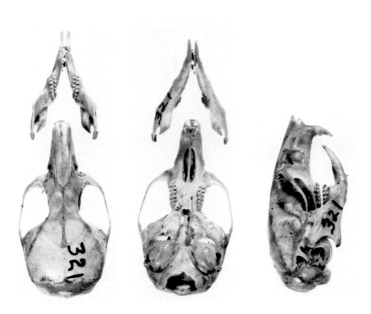

Skull of *Peromyscus maniculatus*, x 2.1

The nest is rather bulky for so small an animal: up to 25 cm in diameter. The outside is composed of coarse plant material and contains a single entrance. The cavity is lined with soft material, often consisting of thistledown, milkweed down, or feathers.

When frightened from the nest a prairie deer mouse seems to move out quickly, but it is really rather slow; it can run only at a rate of about 9 km per hour. Thus, if it cannot find ready cover, it is easily captured by hand.

These mice are active year-round. The best time to find sign of them is after a snow, when their tracks show that they have moved about, from one clump of cover to another. They range widely, an individual often covering a hectare or more. Unlike harvest mice, prairie deer mice are generally solitary in their nesting habits, but the foraging ranges of many individuals may overlap.

There has been no study of the breeding habits of this mouse in Kentucky. In mild winters they probably breed throughout the year in the western part of the state. Litter size is 2 to 9; 3 to 6 being the usual number. In captivity as many as 10 litters have been born to one female in a year. The gestation period is 22–27 days, but in lactating females it sometimes extends to 35 days. The young are naked and blind at birth. Their eyes open at 14 days, and the babies are weaned when they are 25 days old. At 46 to 51 days of age the females become sexually mature.

Food of the prairie deer mouse consists mainly of seeds. Insects are eaten when available, and some fruit on the ground may be used. Grain scattered on the ground at harvest time is favored in winter, and exceptionally high populations can sometimes be found in grain fields where good cover is available. Some food is stored in burrows by these mice.

The little prairie deer mouse is a food staple of feral cats and of weasels, foxes, and owls. This animal is of little economic importance in Kentucky. Its food is mostly weed seeds, and the little grain and fruit eaten is mostly waste.

Cloudland Deer Mouse   PLATE 18

Peromyscus maniculatus nubiterrae Rhoads

Recognition: Total length 157–201 mm; tail 80–106 mm; hind foot 20–23 mm; ear 17–20 mm; weight 13–29 g. A big-eyed mouse with a long, bicolored tail, big, white-edged ears, a Roman nose, and large hind feet. Color above brown to gray (juveniles); white below.

Variation: There are two subspecies of *P. maniculatus* in Kentucky. The two are so different in appearance, behavior, and habitat that we are treating them in separate accounts. The skulls of the two subspecies, however, are very similar.

Confusing Species: Very similar to the white-footed mouse, *P. leucopus.* They can usually be separated by tail characteristics: in *P. m. nubiterrae* the tail is usually more than half the total length, is 80 mm or more long, is sharply bicolored, and usually has a tiny tuft of hair at the end. Within the limited range of the cloudland deer mouse in Kentucky, *P. leucopus* is the only species with which it could be confused.

▓ *P. m. nubiterrae*

Distribution of *Peromyscus maniculatus* in the contiguous United States

Kentucky Distribution: Known only from Big Black Mountain, in Harlan County; abundant. Ranges from the mountain top down to 820 m on the north slope.

Life History: This is a mouse of the higher elevations of the Appalachian Mountains. On the higher parts of Big Black Mountain it is widely distributed, being found in all habitats where cover is available. Farther down the mountain it is restricted to the dense forests of birch, beech, and maple, where the forest floor remains cool and moist in summer. Boulders, fallen logs, and mossy streambanks are favored habitat. Wherever found, this mouse usually is abundant; often it is the most common mammal.

Cloudland deer mice accept a wide variety of shelters. The nest may be placed in an old chipmunk burrow, a rock crevice, a hollow in a log or stump, or a cabin in the woods. Unlike prairie deer mice, they do not hesitate to climb. A pair was once driven out of a woodpecker hole 12 m above the ground in a dead chestnut tree on Big Black Mountain. Traps set in trees for flying squirrels often take deer mice.

Like other members of the genus, this mouse is nocturnal and is active throughout the year. In winter its tracks can be seen linking the fallen logs, rocks, and brush piles where it forages at night.

Food consists of a wide variety of seeds, nuts, and fruits, as well as many kinds of insects and other animal matter. Some seeds are stored for the winter in hollows of logs or trees.

An old mountaineer who was watching one of us prepare specimens of this animal on Big Black Mountain remarked that these were the mice "that make the mice herds." A little discreet questioning revealed that to him a hoard was a "herd." It seems that before the chestnut trees died out he and his fellows would go into the mountains in late fall or early winter, break open hollow trees, and collect the chestnuts stored there by the mice—sometimes a bushel or more per cache.

The nest is a loose, bulky structure of grass, leaves, paper, or any other such material available. It is lined with finely shredded plant material or feathers.

In Kentucky the breeding season extends at least from June through August. The gestation period is 25 to 27 days. Litter size usually is 2 to 4. Blind and naked at birth, the young open their eyes at about 14 to 15 days of age and are weaned when a month old. A female produces several litters in a season.

Remarks: One summer several cloudland deer mice spent a large part of each moonlit night romping on the roof of our tent on Big Black Mountain. There was nothing there to eat; to judge by the pattering feet, they were simply playing in the moonlight.

Like other members of the genus, these big-eyed deer mice make interesting pets. They are clean and tidy and not unfriendly. They will breed in captivity, and they adapt well to life in a cage. They seem to enjoy an exercise wheel and will keep one going most of the night.

White-footed Mouse;
Wood Mouse   PLATE 19

Peromyscus leucopus (Rafinesque)

Recognition: Total length 156–205 mm; tail 63–97 mm; hind foot 19–22 mm; ear 16–18 mm; weight 17–30 g. Upperparts vary from rich reddish-brown to dark brown (slate gray in juveniles); underparts white.

Variation: Two subspecies are recognized in Kentucky. The small, dark southern race, *P. l. leucopus* (Rafinesque), occurs in the western half of the state. In the eastern half of Kentucky *P. l. novaboracensis* (Fisher) occurs; it averages larger and paler than the southern race. The 2 races intergrade in the Bluegrass region. Specimens from the upland region about

White-footed mouse, *Peromyscus leucopus*

Lexington are referable to *P. l. leucopus*; those from the bluffs and woods along the Kentucky River are *P. l. novaboracensis.*

Confusing Species: The several members of the genus *Peromyscus* that occur in Kentucky can be confusing. Although most specimens are rather easily identified, some individuals are difficult, even for the expert. In Kentucky the range of *P. leucopus* broadly overlaps that of *P. maniculatus bairdii.* These animals can nearly always be separated by the length of the tail, which is greater than 60 mm in *P. leucopus* and less than 60 mm in *P. m. bairdii.*

The golden mouse, *Ochrotomys nuttalli,* is a more uniform golden color than any *Peromyscus,* but brightly colored *Peromyscus* are sometimes mistaken for it. Golden mice can be recognized by their lack of a distinct line of separation between the side and belly coloration and by their somewhat cream-colored underparts, in contrast with the pure white of *Peromyscus.*

On Big Black Mountain where *P. m. nubiterrae* is found, it can be confused with *P. leucopus,* which also occurs there.

Most specimens can be separated by the fact that the tail of the cloudland deer mouse is more than half the total length of the animal, whereas in *P. leucopus* the tail is nearly always less than half the total length. In *P. leucopus* the tail is normally less sharply bicolored, and it lacks the tuft of hair that is usually present on the tip of the tail of *P. m. nubiterrae*. In western Kentucky *P. leucopus* can be confused with the cotton mouse, *P. gossypinus*. Cotton mice average larger and darker than *P. leucopus*, and the hind foot usually is noticeably larger. We have seen a few specimens we are not able to identify, and we suspect occasional hybridization.

Kentucky Distribution: Statewide, from the highest point in the state to the banks of lowland streams in western Kentucky. This is one of our most widely distributed and abundant mammals.

Life History: Wherever adequate cover is available in Kentucky, in the form of trees or brush, you can expect to find the white-footed mouse. Even in the cities, if parks and streambanks are not too manicured, these mice can sometimes be found.

Woodlands with plenty of cover provided by brush piles, fallen logs, and boulders are the preferred habitat. Cliffs and caves are also popular with these mice, and man-made structures are readily used. A summer camp or cabin in the woods is occupied by these mice at the first opportunity.

Brushy fencerows and open woodlands with small trees or with shrubs, such as blackberries, also provide suitable habitat for these mice. However, extensive weedfields and croplands not bordered by trees or shrubs are not suitable for this species; such sites are occupied by the prairie deer mouse, *P. maniculatus bairdii*. Some areas that are intermediate—those containing brush and weeds—are occupied by both species.

White-footed mice are nocturnal and are active throughout the year. Even on the coldest nights they scurry about on the forest floor, leaving their tracks in the snow. They do not make

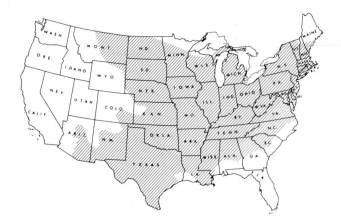

Distribution of *Peromyscus leucopus* in the United States

Skull of *Peromyscus leucopus*, x 2.1

runways, as some of our small mammals do; therefore no paths are evident when there is no snow cover. However, trails in the snow indicate that these mice have regular patterns of movement from one place of shelter to another.

A wide variety of nest sites seems suitable for this animal. Nests are commonly built above ground, in hollow trees, woodpecker holes, or birdhouses. Often an abandoned nest of a bird is roofed over and used, or a leaf nest made by a squirrel is occupied. Nests are often located on the ground, beneath a stone or a pile of trash. Cliffs, caves, rock crevices, and buildings also serve as nest sites. Although these mice do not dig burrows, they will use the burrows of chipmunks or other animals and sometimes nest underground.

The nest is a globular structure, usually about 150 to 250 mm in diameter, and is composed of just about any soft material available: grass, leaves, paper, feathers, or cloth.

After a gestation period of about 23 to 25 days the first litter is born in April; this is followed by 3 or more litters by November. Litter size varies from 1 to 7; the usual number is 3 to 6. An older mouse tends to have larger litters than a young mouse with her first litter. Females can breed when only 10 weeks old.

At birth the mice weigh about 1.5 to 2 g and are blind and hairless. Their eyes open in 12 to 14 days, and the young are weaned when about 24 to 28 days old. After they leave the nest the young can be recognized by their uniformly gray upperparts. As they begin to molt to adult pelage, brown comes in on the sides, giving a peculiar tricolored pattern of gray, brown, and white. As the molt proceeds, the last region to lose the gray is the top of the head.

The white-footed mouse feeds on a wide variety of seeds, nuts, and grain. In Kentucky wild cherry seeds seem to be a favorite, and piles of opened seeds are often seen in feeding areas frequented by these mice. Acorns and hickory nuts are eaten, as are many tiny seeds of grasses, clover, and various weeds. Insects and other invertebrates supplement the varied diet. When seeds are abundant, they are gathered and stored in cavities, where they will be available in winter.

Owls, weasels, foxes, cats, snakes, and other animals feed heavily on white-footed mice. The animal is abundant enough to be a major food item of predators.

White-footed mice seldom are of much economic significance to man. They occasionally invade the grain bin and do some damage, and they can be a nuisance when they take up residence in buildings. They are very attractive little animals and make delightful pets.

Cotton Mouse   PLATE 19

Peromyscus gossypinus (Le Conte)

Recognition: Total length 160–205 mm; tail 68–97 mm; hind foot 21–26 mm. Upperparts dark brown (gray in juveniles); underparts white. This is the large, dark, white-footed mouse of the Deep South.

Variation: No geographical variation is evident in Kentucky. The subspecies here is *P. g megacephalus* (Rhoads).

Confusing Species: This species is easily separated from the prairie deer mouse, *P. maniculatus bairdii,* by its longer tail, and from the golden mouse, *Ochrotomys nuttalli,* by its darker

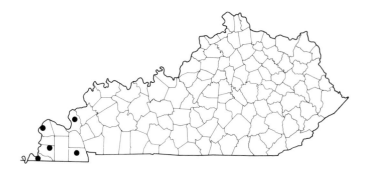

Locality records of *Peromyscus gossypinus* in Kentucky

color. It is most similar to the white-footed mouse, *P. leucopus*. Most specimens can be separated from the latter on the basis of their larger size, larger hind foot (usually 22 mm or more), and larger skull (28 mm or more). However, we have seen specimens we could not identify, and we suspect occasional hybridization occurs in Kentucky, as it does in other parts of the range.

Kentucky Distribution: Western Kentucky; apparently limited to the Purchase. Fairly common in suitable habitat.

Life History: In Kentucky this is an animal of the wooded streambanks, swampy woods, and brushland of the Purchase. Most of the habitat of the cotton mouse is subject to annual flooding. *P. gossypinus* tends to prefer the lowlands, whereas *P. leucopus* tends to select the higher, better-drained woods and brushland; however, there is some overlap of habitat between the 2 species.

Like other species of *Peromyscus*, the cotton mouse is nocturnal and active throughout the year. It is partial to areas with good cover, such as fallen logs, brush piles, and old buildings. Nest sites tend to be on the higher spots or located above ground, in hollow stumps or trees.

The breeding season in Kentucky is unknown. In Florida cotton mice breed throughout the year, except for midsummer. The gestation period is 23 days for non-nursing females and 30 days for those that are lactating. Litter size averages 4 and ranges from 1 to 7. The newborn young are pink and hairless. Their skin is wrinkled, their eyes are closed, and their ears are folded down over the openings. The ears unfold on the fourth day, and hair begins to appear on the fifth day. The eyes open in about 2 weeks. By 3 weeks of age the young are able to forage for themselves, and nearly all are weaned by the age of 1 month. In the laboratory, young have been separated from the mother when their eyes opened and then successfully weaned.

As with the other species of *Peromyscus*, the juvenile dorsal

Distribution of *Peromyscus gossypinus*

Skull of *Peromyscus gossypinus*, x 1.8

pelage is a uniform gray. Molting to adult pelage begins between the 32nd and 53rd day. The first sign of molt is a brown band of adult pelage on each side.

Cotton mice are more carnivorous than our other species of *Peromyscus*. A study at Reelfoot Lake, Tennessee, found that 68% of their food was animal matter, consisting mostly of beetles, moth larvae, spiders, and slugs. Most of the plant matter eaten consisted of spores of the fungus *Endogone*.

Like other species of white-footed mice, cotton mice take well to captivity. They breed readily and raise young in capitivity.

The cotton mouse is of little economic importance in Kentucky, but it occasionally enters buildings and becomes a minor nuisance.

Golden Mouse   PLATE 20

Ochrotomys nuttalli (Harlan)

Recognition: Total length 150–190 mm; tail 68–93 mm; hind foot 17–20 mm. A richly golden-colored mouse. Upperparts nearly a uniform golden brown; underparts pale cream-colored, without a clear line of separation between back and belly.

Variation: Two subspecies occur in Kentucky. *O. n. aureolus* (Audubon and Bachman) is found across eastern and central parts of the state. In the Purchase we find a smaller race, *O. n. lisae* Packard.

Confusing Species: The only other mouse that is so brightly colored is the woodland jumping mouse, *Napaeozapus insignis*, which is easily distinguished from the golden mouse by its much longer tail. Brightly colored individuals of the various species of *Peromyscus* are occasionally mistaken for golden mice, but they all have a sharp line of separation between the back and belly fur.

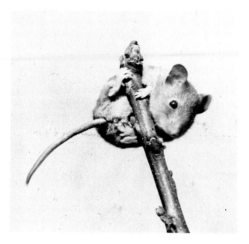

Golden mouse,
Ochrotomys nuttalli

Nest of a golden mouse

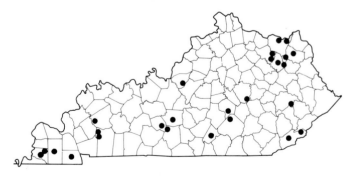

Locality records of *Ochrotomys nuttalli* in Kentucky

Kentucky Distribution: Statewide, except for the Inner Blue-grass and perhaps northern Kentucky.

Life History: This is an animal of the greenbrier (*Smilax*) thickets. Woodlands with a dense understory of greenbrier, some of which is growing up into the trees and shrubs, are the favorite habitat. Pines and cedars are also especially favored. Other plants that sometimes provide suitable habitat for golden mice in Kentucky include honeysuckle, blackberry, and grapevine.

In winter, when the leaves are off, one can rather easily locate golden mice by searching out their arboreal nests. These spherical structures, about 125 mm in diameter, are usually within reach but occasionally as high as 9 m from the ground. Golden mice can be captured by creeping up to the nest and placing a hand over the entrance. The mice are sociable little creatures, as many as 6 adults sometimes occupying a single nest. The outer layer of the nest consists of leaves, and the inner chamber is lined with shredded bark or grasses. Feathers are often used. Sometimes an old bird nest is roofed over and used by golden mice.

In addition to the nests, in which the mice spend the day, there are numerous feeding platforms in trees or vines; often they are deserted bird nests or abandoned mouse nests. The mice rest and feed on these platforms, which, typically, are covered with seed hulls and give a good index of the feeding habits of the mice. Seeds of greenbrier, wild cherry, sumac, dogwood, and blackberry are among the favorite foods. Some insects are also eaten.

The home range is rather small: a little over 0.2 hectare. In good golden mouse habitat the ranges of a group living in one nest may overlap the ranges of their neighbors, and they may share feeding platforms.

The breeding season in Kentucky extends at least from March through October. The gestation period is 25 to 30 days. The litter size of 1 to 4, usually 2 or 3, is smaller than in most mice. However, because the female is sometimes

Distribution of *Ochrotomys nuttalli*

Skull of *Ochrotomys nuttalli*, x 1.9

impregnated the day after giving birth, several litters can be raised in a season.

The newborn young are hairless, with closed ears and eyes. The external ear unfolds and becomes erect at 2 or 3 days of age, and the eyes open at about 13 days. Weaning starts when the young are 17 or 18 days old; in their fourth week the young are able to forage for themselves.

Golden mice are active only at night, foraging both on the ground and in the trees and vines. Seeds gathered are stored in cheek pouches and carried to a feeding platform. The mice are very competent above ground, scurrying deftly along the smallest branches with the aid of their semiprehensile tails. They are active year-round.

Remarks: Golden mice are among the most docile of our wild animals and make interesting pets. Their gentle disposition and striking color make them one of our most attractive native animals. They never become pests or do any economic damage, as far as we know.

Hispid Cotton Rat   PLATE 20

Sigmodon hispidus (Say and Ord)

Recognition: Total length 224–292 mm; tail 81–124 mm; hind foot 28–34 mm; ear 16–24 mm; weight 54–158 g. A robust rat with coarse, grizzled pelage. Upperparts dark, varying from brown to nearly black; underparts grayish-white or buff.

Variation: No geographic variation is evident in Kentucky. Our subspecies is *S. h. hispidus* (Say and Ord).

Confusing Species: Most similar to the much smaller prairie vole, *Microtus ochrogaster*, with which it is often associated. The cotton rat can be recognized by its much larger size and longer tail. Its heavy body and relatively shorter tail distinguish it from the slim, long-tailed rice rat, *Oryzomys palustris*.

Kentucky Distribution: Western Kentucky. The cotton rat is a southern animal that has been steadily extending its range northward in recent years. The first Kentucky specimens were taken in Lyon County in 1964. The species now seems to be well established and fairly common in the Purchase and the Land Between the Lakes.

Life History: The cotton rat is an animal of the weedy roadside ditches and fencerows of western Kentucky. It is an abundant animal, being found wherever adequate cover is available.

The presence of cotton rats is easily detected by looking for their sign. Broad runways, about 75 mm in diameter, lace the grass wherever these animals are found. Their droppings are distinctive: oval and about 3 mm in diameter, in contrast to the long, slim droppings of mice, about 1.5 mm across, and the larger, spherical droppings of rabbits.

Cotton rats are frequently seen foraging by day and are are easily caught in traps set in runways. Their nests usually are in shallow burrows but are sometimes placed in a thick clump of grass or a brush pile. The home range usually is less than half a hectare—rather small for an animal of this size.

A wide variety of food is used: leaves, stems, roots, and seeds of many common plants, as well as animal matter, including insects, crayfish, and the eggs and young of ground-nesting birds. A high population of cotton rats can be rather destructive to bobwhite quail.

In the Deep South cotton rats can be a major pest to farmers. They have been reported especially destructive to sugar cane, cotton, sweet potatoes, and squash. We have no reports of their being a nuisance in Kentucky.

Cotton rats are among the most prolific of mammals. After a gestation period of 27 days a litter of 2 to 10 (average 5) is produced. The young are blind but are haired and able to run. Their eyes open within 2 days, and the young may be

Distribution of *Sigmodon hispidus* in the United States

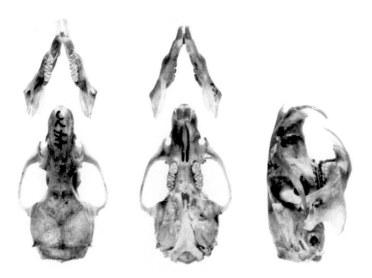

Skull of *Sigmodon hispidus*, x 1.3

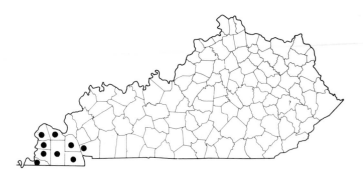

Locality records of *Sigmodon hispidus* in Kentucky

weaned when only 5 days old. When 40 to 50 days old both sexes are ready to breed. One female raised in captivity produced a litter when she was 67 days old.

In captive animals copulation often occurs within 3 to 6 hours after a litter is born, and the mother ovulates at 6 to 12 hours after giving birth. With the breeding season lasting most of the year, the reproductive potential is enormous. One captive female produced 9 litters in 10 months, and many produced 6 or more litters in this time-span.

Remarks: Cotton rats have been adapted to captivity and are used as experimental animals in various types of biological and medical research.

Thane Robinson, of the University of Louisville, collected a cotton rat near Quicksand, in Breathitt County. This locality is well out of the known or expected range of the species, and we are at a loss to explain its presence there.

Eastern Woodrat;
Cave Rat; Pack Rat   P<small>LATE</small> 21
Neotoma floridana (Ord)

Recognition: Total length 320–477 mm; tail 140–210 mm; hind foot 33–50 mm; weight 394–609 g. This is a large, handsome gray rat with white feet and belly. The tail is haired and is bicolored: gray above and white below.

Variation: Two subspecies occur in Kentucky. From Mammoth Cave eastward we find the larger, gray *N. f. magister* Baird. In the Purchase is *N. f. illinoensis* Howell, a smaller race with more brownish to yellowish fur.

Confusing Species: This large native rat is not likely to be confused with any other mammal in Kentucky except the Norway rat, *Rattus norvegicus.* The latter can be recognized by its brown color and naked, scaly tail.

Kentucky Distribution: Throughout most of the state wherever there are caves, cliffs, rocky outcrops, or talus slopes. Woodrats are particularly common in the sandstone cliffs of the Cumberland Plateau. They are absent from the Inner Bluegrass northward but occur in the palisades along the Kentucky River. West of Mammoth Cave woodrats are scarce and local (as they are in southern Indiana and Illinois). They are also scarce and local in the Purchase, where they live in swamps and abandoned buildings.

Life History: The woodrat is associated with rocky outcrops. Cliffs with deep crevices, caves, or large boulders piled in such a way as to form numerous retreats and shelters are favored by these beautiful animals.

The presence of woodrats is easy to detect. Wherever they are found, they make distinctive piles of sticks in crevices and on rock ledges. These piles often contain assortments of any kind of trash available to the animals, such as flash bulbs, shotgun shells, bits of glass and paper, bones, metal scraps, and articles of clothing. Also included are cut pieces of green vegetation, such as cedar, sumac, and rhododendron.

Farther back in a crevice, in a dark, more protected area, one may find the woodrat nest. This neat structure, usually about 3 to 5 dm in diameter, is constructed of shredded bark.

Distribution of *Neotoma floridana*

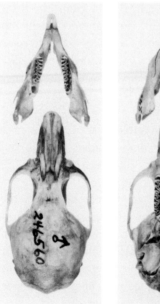

Skull of *Neotoma floridana*, x 1.3

The nest chamber, about 100 mm in diameter, is open, giving the structure the appearance of a giant bird nest. Here the woodrat spends the day. When approached with a light in the darkness of a cave, the rat is remarkably tame, sometimes allowing one to get within arm's reach.

Woodrats are active by night and at all seasons. They forage out away from the rock cliffs and feed upon a wide variety of plant material, including fruits, berries, and green vegetation. In fall they hoard piles of acorns and other seeds. Apparently they eat little animal matter. In captivity they take most animal matter reluctantly but are quite fond of the oil in which sardines are packed.

These animals clamber over the face of a cliff or up the wall of a cave with ease. When foraging in trees and shrubbery they are nearly as agile as squirrels, climbing out to the end of a branch to cut twigs for the midden back home.

After a gestation period of 30 to 36 days a litter of 1 to 3 young is born, in March. Probably several litters are produced each year; young in the nest have been found as late as September.

The young weigh about 15 g and are naked and helpless at birth. They mature slowly; their eyes open about the 19th day. By the 24th day they begin to forage away from the nest and can eat some solid food.

Woodrats are remarkably gentle animals. Those we have

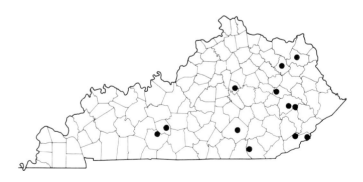

Locality records of *Neotoma floridana* in Kentucky

handled made no effort to bite, even when just captured in the wild. They readily breed and raise young in captivity.

Remarks: In Kentucky this species seldom comes into contact with man and is considered of little economic importance. Although it is nearly the size of a gray squirrel and the flesh is reported to be delicious, there is little prospect of its becoming a game animal. It is to be valued as an interesting member of our natural fauna—a close relative of the legendary pack rat of the West.

The distribution of woodrats in Kentucky is not thoroughly understood. Collecting efforts should be directed toward determining whether they occur in the caves and cliffs of Trigg, Livingston, and Crittenden counties, in western Kentucky.

Gapper's Red-backed Mouse   PLATE 21

Clethrionomys gapperi (Vigors)

Recognition: Total length 125–172 mm; tail 33–49 mm; hind foot 19–20 mm; weight 16–42 g. This is a medium-sized vole with relatively large ears and small eyes. It is characterized by a broad dorsal band of rusty or reddish fur. The sides are brown; the underparts buffy white to pale gray.

Variation: The subspecies in Kentucky is *C. g. maurus* Kellogg, which was described from specimens taken on Big Black Mountain. It is a dark race; in the darkest individuals the dorsal band is so obscure that the animal, unless examined carefully, might be mistaken for *Microtus*. The brightest individuals are recognizable at a glance as red-backed mice.

Confusing Species: The only other microtines that have been found on Big Black Mountain, *Synaptomys cooperi* and *Microtus pinetorum*, are easily recognized by their shorter tails.

Habitat of the red-backed mouse near the summit of Big Black Mountain, Harlan County, Kentucky

Juvenile red-backed mice and the darkest adults are very similar in general appearance to M. *pennsylvanicus*.

Kentucky Distribution: Known only from Big Black Mountain, in Harlan County, where it ranges from the summit down to about 680 m elevation.

Life History: This is a northern species that reaches the southern limits of its eastern range in the southern Appalachians. Its favored habitat on Big Black Mountain is the cool, damp woodlands, especially where the microhabitat is protected by logs and rock piles. However, at the higher elevations red-backed mice are generally distributed; high on Big Black Mountain we have taken them in dry, bushy areas where the dominant vegetation is hawthorne and rhododendron; in oak-hickory woods; and among the tulip trees in a dry, rocky area.

This vole is more agile than other microtines. It scampers about on the ground, along logs and over rocks, foraging for seeds and other vegetable matter. It is primarily nocturnal but is sometimes active by day.

Red-backed mice live in burrows. They sometimes use runways made by moles, chipmunks, or short-tailed shrews, but they are much less confined to runways than other microtines;

Distribution of *Clethrionomys gapperi* in the contiguous United States.

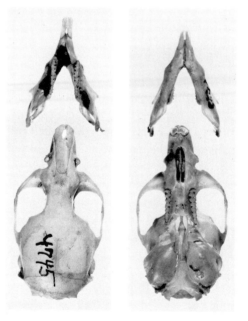

Skull of *Clethrionomys gapperi*, x 2

they spend much of their active time on the open forest floor. Cuttings of ferns accompanied by small black droppings similar to those of the house mouse betray the presence of these mice in the mountain forests.

The few nests described have been found in shallow burrows as much as 450 mm beneath the ground. Nests are 75–100 mm in diameter and consist of grass, leaves, and moss.

Breeding begins in March or April and continues until October. Litter size in Kentucky is usually 3 or 4. The young are born blind and hairless, and they weigh about 1.9 g each. Their eyes open in 12 or 13 days, and the young are weaned at 17 days of age.

Red-backed mice are active in the coldest weather, and their tracks in the snow make well-marked paths through their favored foraging areas, such as fallen trees, brush piles, and rock piles. The home range averages about 0.1 hectare.

Although the agile red-backed mice are not as easily captured by owls as their lumbering relatives the meadow voles, they are taken by a wide variety of predators. Rattlesnakes commonly eat them and may be their most important natural enemies.

Red-backed mice seldom come into contact with man and are of little economic importance. Their diet of wild berries, seeds, fungi, ferns, and an occasional insect is of little concern to us. These mice are of value as a very local, unique part of our native fauna.

Meadow Vole; Meadow Mouse;
Field Mouse   PLATE 22

Microtus pennsylvanicus (Ord)

Recognition: Total length 140–195 mm; tail 32–64 mm; hind foot 18–24 mm; weight 31–43 g. This is a large, robust, relatively long-tailed vole. Color above is dull chestnut brown,

Distribution of *Microtus pennsylvanicus* in the contiguous United States

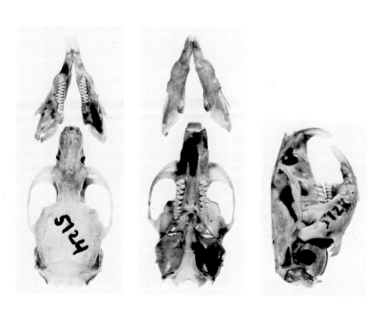

Skull of *Microtus pennsylvanicus*, x 1.7

varying from dark brown to bright chestnut; underparts silvery gray.

Variation: We see no geographic variation among Kentucky specimens. Our subspecies is *M. p. pennsylvanicus* (Ord).

Confusing Species: There are 4 other species of short-tailed mice in Kentucky. *M. pinetorum* and *Synaptomys cooperi* have much shorter tails than the meadow vole. *Clethrionomys gapperi* usually has a broad reddish band down the back; however, in a few dark individuals this band is indistinct, and they look much like young meadow voles. In Kentucky, however, these 2 species do not occur together.

The meadow vole is most similar to the prairie vole, *M. ochrogaster.* The latter can be distinguished by its somewhat shorter tail and the grizzled appearance of the dorsal fur. Meadow voles always have silvery belly fur. Prairie voles from central and western Kentucky have a wash of yellow on the belly fur; those from eastern Kentucky and the Cumberland Plateau are white-bellied.

Kentucky Distribution: Abundant from the Bluegrass of central Kentucky northward and eastward. To the west and south it becomes rare and local, extending little beyond the limits of the Knobs.

Life History: This is an animal of the grasslands and open swamps. It is a northern species, reaching the limits of its range in the southern Appalachians and in central Kentucky. In northern climates, where meadow voles are abundant, they can be found in any roadside ditch, fencerow, or other grassy spot large enough to provide sufficient cover.

In Kentucky this species competes with the prairie vole for the grassland habitat available. Around Lexington, where both species are common and the ranges of individuals of each species often overlap, we can discern a difference in habitat preference: *M. pennsylvanicus* prefers the more moist lower

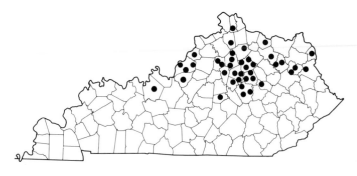

Locality records of *Microtus pennsylvanicus* in Kentucky

grasslands, and *M. ochrogaster* prefers the uplands. *M. pennsylvanicus* likes swamps and does not hesitate to enter the water.

Meadow voles betray their presence with conspicuous signs. Beneath the matted grass are well-worn runways about 40 mm in diameter. Piles of cut grass stems and mouse droppings are also evidence of their presence. The runways are used so regularly that the easiest way to catch a meadow vole is to set a trap across a well-worn run.

Meadow voles are active day and night and throughout the year. Activity is greatest in the morning and late afternoon, as the voles go about their daily work of foraging for food and keeping the runways clean.

Food consists chiefly of grass and other herbage. A wide variety of vegetable matter, including grain, fruit, and alfalfa, is eaten; so is animal matter, when available. The mice are sometimes cannibalistic, killing and devouring young if confined together.

The home range of meadow voles has been studied by many investigators, and the findings vary widely. One reported the range as less than 270 m². Others found that these voles commonly range over nearly half a hectare. The range is smallest when the mice are most abundant. However, they are not territorial; the ranges of several individuals overlap broadly.

Like most small mammals, meadow voles return home after displacement. One researcher found that, when conveyed to

Nest of meadow vole undisturbed (left), and torn open

points 335 to 366 m from their homes, about a third of them returned successfully, but they were unable to return from 430 m or farther.

The nest is usually built on the surface of the ground, in a tussock of grass, but occasionally a nest is made in a shallow burrow or under a board or a piece of trash. The nest is spherical and about 150 mm in diameter. It consists of dry grass; the inner lining is finely shredded and soft.

Concerning reproduction, W. J. Hamilton, Jr., writes that of all prolific mammals the meadow vole is the champion. One litter follows another in rapid succession, because the mother is impregnated anew as soon as a litter is born. One captive female produced 17 litters in a year, and one of her daughters produced 13 litters before she had reached her first birthday.

The gestation period is 21 days. Litter size, ranging from 1 to 10, is commonly 6 or 7. At birth the young are naked and blind and weigh about 3 g. They grow rapidly, gaining about 1 g a day until half-grown. Their eyes open about the eighth day, after which they begin to wander outside the nest. At 12 days of age they are weaned, and when 25 days old the females are ready to breed. The breeding season is year-long. In northern regions meadow voles breed beneath the snow throughout the winter.

The population of meadow voles is subject to dramatic

fluctuations. Some years they seem literally to overrun the countryside, with populations in favorable habitats reaching 500 or more per hectare. Following such outbreaks the population crashes, and the animals become quite scarce and difficult to find in the following year. The population gradually builds back up until another crash occurs.

Meadow voles are sometimes of considerable economic significance in Kentucky. In winter they may girdle small fruit trees that are unprotected. High populations of voles can affect the productivity of hayfields. They sometimes invade grain fields and eat both foliage and grain.

Just as meadow voles have a high breeding potential, they also have a high rate of loss to predation. They form the mainstay of the diet of many predators and are taken by nearly all predatory species of birds and mammals. Screech owls, barn owls, sparrow hawks, and red-tailed hawks take them in large numbers; so do weasels, mink, and foxes. House cats find the large, bumbling meadow voles much easier to catch than other species, and capture them out of proportion to their population.

Black racers, rat snakes, and even the common little garter snakes feed upon meadow voles. We are fortunate to have such a diversity of wild predators to keep the population of these interesting little rodents in check.

Prairie Vole   PLATE 22

Microtus ochrogaster (Wagner)

Recognition: Total length 130–172 mm; tail 24–41 mm; hind foot 17–22 mm; ear 11–15 mm; weight 37–48 g. A dark, short-tailed vole with long, coarse pelage. Upperparts usually grizzled gray, but varying from chestnut brown to nearly black; underparts whitish or washed with buff or yellow.

Variation: Two subspecies occur in Kentucky. From about Lexington westward *M. o. ochrogaster* (Wagner) occurs; this

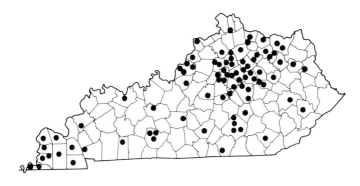

Locality records of *Microtus ochrogaster* in Kentucky

is a large, dark vole with a buffy belly. Eastward occurs *M. o. ohionensis* Bole and Moulthrop; it is more brown, has silvery-white underparts, lacking the buffy tips to the belly fur.

Confusing Species: The pine vole, *M. pinetorum*, and the southern bog lemming, *Synaptomys cooperi*, are easily identified by their shorter tails. The meadow vole, *M. pennsylvanicus*, has a somewhat longer tail and never has a buffy belly. Some individuals of this species are quite similar to *M. ochrogaster* and can be identified with certainty only by the tooth pattern (see key).

Kentucky Distribution: Nearly statewide; abundant. Absent from the southeastern mountains and adjacent areas. This species has been extending its range eastward with the pasture grasses, particularly fescue, and is now found in the hollows of many mountain counties, at least to Knott County.

Life History: This is the common vole in Kentucky, found nearly everywhere there is sufficient grass for cover. In parts of the state where the meadow vole is absent, this species occupies all suitable habitat, including cattail swamps. However, where the 2 species occur together, the prairie vole selects the more upland, drier habitat. Its presence is easily detected by the network of runways, as well as grass cuttings and slim black droppings.

Excavated nest of prairie vole

This species is more fossorial than the meadow vole, tending to nest in burrows. Thus, it occupies areas where cover is sparser than would be suitable for the meadow vole. In a city park in Lexington we once found prairie voles living in burrows on a hilltop that was mowed as closely as a lawn. On the campus of the University of Illinois we have seen them living in a manicured rock garden where cover was scarcely suitable for any other native small mammal.

On grassy slopes near Lexington we have observed that the nesting sites of prairie voles can be located at a glance. The vegetation about the site is lush and greener, apparently as a result of the nitrate provided by the droppings.

The bulky nest, of dried grass, is placed in an underground chamber. Here the young are born in the course of a breeding season that may last throughout the year. Not nearly as prolific as meadow voles, prairie voles have litters ranging from 1 to 9; the average is 3.4 in the wild and 4.6 in laboratory colonies. Gestation is about 21 days. A female produces 3 or 4 litters per year, and the young mice breed when about 30 days old.

Prairie voles eat grass, stems, roots, tubers, seeds, and fruit, and they may strip and eat the bark of shrubs and small trees. Unlike the meadow vole, this vole sometimes stores food. Underground chambers are sometimes stuffed with clippings of bluegrass or alfalfa. In a burrow in Kansas 7 liters of seeds from Kentucky coffee trees were once found, and there is a

Distribution of *Microtus ochrogaster* in the United States

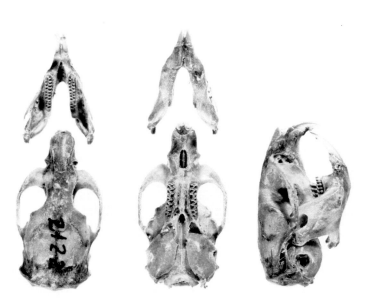

Skull of *Microtus ochrogaster*, x 1.8

report of prairie voles storing 4 liters of the fruits of the house nettle.

As with other voles, the population of this species is quite variable. In Kentucky the population seems to be correlated with weather patterns. During a succession of extremely dry years, in the 1960s, this species (as well as meadow voles and lemming mice) became so scarce around Lexington that searchers of prime habitat sometimes failed to turn up any sign of them whatsoever.

Prairie voles have very small home ranges. Studies at Lexington of animals marked with radioactive tags showed home ranges that varied from 80 to 725 m². Several voles occupied the same general area, with home ranges overlapping.

Prairie voles are competent at finding their way home over short distances. When displaced 100 meters or so from home in the daytime, a vole goes beneath the grass cover and rests until dusk, when it emerges and goes straight home.

These voles are active at all daylight hours, but at night they emerge from the underground nests only infrequently and for only a few minutes at a time. On the coldest days in winter they are most active at midday; at other seasons they are most active in the morning and late afternoon.

This species is of considerable economic importance in Kentucky. When the population is high, prairie voles can eat significant amounts of hay and grain and can destroy young fruit trees by eating the bark. They are a staple food of a wide variety of native predatory birds, mammals, and snakes; these are responsible for keeping the population of voles under control.

Pine Vole; Pine Mouse;
Woodland Vole   PLATE 23

Microtus pinetorum (Le Conte)

Recognition: Total length 105–145 mm; tail 17–25 mm; hind foot 15–20 mm; ear 8–12 mm; weight 22–39 g. A short-tailed vole with small eyes and small ears. Color brown, varying from nearly black to bright russet. Lower parts dusky to silvery gray. The fur is soft, short, and dense.

Variation: Two subspecies occur in Kentucky. The darkest race of the pine vole, *M. p. carbonarius* (Handley), was described in 1952 on the basis of specimens from Eubank, and adjacent hilly country Pulaski County, Kentucky. This subspecies ranges across the Cumberland Plateau and throughout the Cumberland Mountains. A slightly paler form, *M. p. auricularis* Bailey, is found in the rest of the state.

Confusing Species: The only other vole in Kentucky with a comparably short tail is the southern bog lemming, *Synaptomys cooperi*. It is readily distinguished by the shallow groove on each upper incisor.

Kentucky Distribution: Statewide. Abundant in the Bluegrass and eastward; somewhat less common in the western half of the state.

Life History: This animal is found in a wide variety of habitats, from woodland to grassland; in fact, it is found almost anywhere there is adequate cover, friable soil, and an adequate food supply. Along fencerows and roadbanks in the Bluegrass this species is often the most abundant small mammal.

Pine voles are burrowing animals that spend most of their time below ground. Burrows are located just under the leaf mold or other cover and may extend several centimeters into the soil. Frequent openings to the surface, sometimes accompanied by a little loose dirt, are characteristic of these voles.

Pine voles are more often trapped at night than in the daytime. However, studies by Faith Hershey at Lexington showed that their activity is random. Apparently they are more active by day below ground and are more likely to appear at the surface after dark.

The home range is small, some individuals apparently living

Distribution of *Microtus pinetorum* in the United States

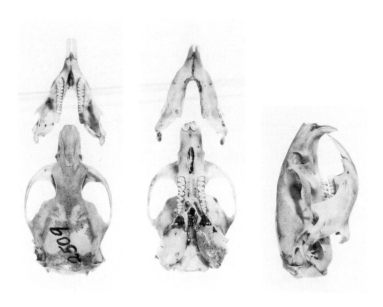

Skull of *Microtus pinetorum*, x 1.7

about the root system of a single tree. As with other species, pine voles that are released a short distance beyond their normal range readily come home.

The reproductive season extends from January through October. One to 7 young, but usually 2 to 4, are born in a nest of leaves, stems, or rootlets under a stump or log or on the surface of the ground. The young are naked and helpless at birth but grow rapidly; their eyes open at age 9 to 12 days. By 2 weeks of age the young voles are wandering outside the nest, and they are weaned toward the end of the third week.

Food consists of a wide variety of plant and animal matter, but roots, tubers, bark, and bulbs are the mainstays; thus, these voles are more destructive to garden crops than any other mouse. Mole runs in the garden are often used by pine voles to get to potato tubers, which they relish.

Orchards are often inhabited by pine voles, and considerable damage results. Whereas our other voles, as well as rabbits, primarily do their damage by eating the bark of young fruit trees in winter, pine voles attack trees of all sizes at all seasons. Most of their damage is below ground, where they eat rootlets and the bark of larger roots. As with other species of mice, the population of pine voles fluctuates greatly; the year following heavy damage, the population is likely to be low, whether control measures are applied or not.

These are among the most cannibalistic of our small mammals. When a pine vole is killed in a trap it is common for another to devour part of it. Sometimes when trapping we are forced to catch most of the pine voles before we can obtain intact specimens of other species. Insects and other animal matter are also readily eaten.

These voles sometimes fight viciously, to the point of injuring one another. Fights are noisy, with harsh chattering and tooth-gnashing. Although some individuals can be handled easily, most of them bite and never become docile.

The running speed of 6 km per hour makes this one of our slowest voles. It falls victim to owls, hawks, foxes, raccoons, opossums, cats, dogs, and mink, among other predators. On Big Black Mountain one was taken from the stomach of a black rat snake. Because of its fossorial habits, however, this vole is less often taken by predators than are our other species.

Pine voles are often heavily infested with parasitic worms, lice, and mites; these may be factors in the crash that occurs when the population gets too high.

Sometimes pine voles seem difficult to catch in traps. Usually, however, traps placed at burrow openings or across well-worn tunnels will take a few.

Muskrat PLATE 23

Ondatra zibethicus (Linnaeus)

Recognition: Total length 409–635 mm; tail 180–295 mm; hind foot 64–88 mm; weight 541–1,814 g. A large, aquatic rat with large, webbed hind feet and a long, scaly, sparsely haired, laterally compressed tail. Ears small and nearly hidden in the fur. The dense, waterproof pelage consists of soft underfur overlaid by long guard hairs. Rich brown above, paler below.

Variation: No geographic variation is recognized in Kentucky. Our subspecies is *O. z. zibethicus* (Linnaeus).

Confusing Species: Muskrats are sometimes mistaken for small beavers, *Castor canadensis,* but beavers are easily recognized by their broad, flat tails.

Kentucky Distribution: Statewide; abundant in suitable habitat.

Life History: This large rat is most at home in the water. In Kentucky its favored habitat is along the slow-moving lowland streams and about farm ponds. Kentucky has little of the marshy habitat so much favored by muskrats in other states.

Distribution of *Ondatra zibethicus* in the contiguous United States

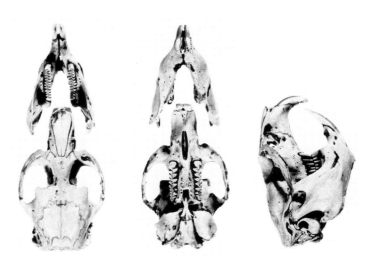

Skull of *Ondatra zibethicus*, x 0.7

In open marshlands, muskrats build houses consisting of a heap of vegetation sometimes a meter high and 3 meters across. Although such houses are scarce in Kentucky, they can be found in a few of our widely scattered marshes and in swampy areas about impoundments.

Most of our muskrats dwell in burrows in the bank of a stream or pond. The entrance is a hole, 200 to 250 mm in diameter, that opens underwater into a lateral burrow. The burrow rises to a dry nest chamber. The top frequently breaks through, so that the bank of a stream or pond occupied by muskrats is often marked with broad holes. The comings and goings of the muskrats commonly cause the formation of an underwater channel leading to the entrance to the burrow.

Muskrats are primarily nocturnal, but they are active enough in the daytime for an alert observer who spends much time outdoors to be familiar with them. They are usually seen swimming across a river or pond, moving quietly, with only the nose above water. If frightened the muskrat will dive. He is a superb swimmer and may travel as much as 55 m underwater without coming up for air. Muskrats are reported to be able to stay underwater for 15 minutes.

Little is known about reproduction in the muskrats of Kentucky. In the northern states the first litter is born in April, and 3 or 4 litters are usually produced in a season. In Louisiana breeding occurs throughout the year.

In Wisconsin the gestation period is about 29 days, and the average litter size is 7 or 8. Litters as small as 1 and as large as 16 have been reported. In Louisiana the litter size usually is 3 or 4. The young are blind, naked, and helpless at birth, and they weigh about 21 g. Pelage develops rapidly: by the end of the first week they are well furred. Their eyes open at 14 to 16 days of age—by which time the agile youngsters can swim and dive. They are weaned at 4 weeks but do not become sexually mature until about 10 to 12 months old.

Muskrats have a voracious appetite, taking about a third of their weight in food every day. Most kinds of aquatic vegetation are eaten. In Kentucky cattails seem to be a favorite;

Muskrat house in a marsh
near Lexington

Well-used bank den
of a muskrat

DELBERT RUST

Swimming muskrat

Muskrat,
Ondatra zibethicus

sometimes the first evidence of muskrats in a farm pond is floating pieces of cattail that have been pulled up and partially eaten. Muskrats rarely forage far from the water, but at times they may raid a nearby cornfield, causing minor damage. In marshland they make well-worn trails throughout the vegetation. Crayfish, clams, frogs, and almost any kind of dead animal matter, including their own kind, are occasionally eaten by hungry muskrats.

As the breeding season approaches, in early spring, many muskrats leave the waterways and travel overland. They may then be seen miles from water; and it is not uncommon to see muskrats dead on the road in Kentucky in the spring. Wandering muskrats also appear on the streets, in even our largest cities, at this season; perhaps they have come up from the storm sewers.

After a successful breeding season the muskrat population is often larger than can be sustained through the winter; this occasions a fall season of wandering, as the surplus animals disperse in search of suitable new habitats. Nearly all of these wandering animals are killed by automobiles and predators.

Mink are the major predators on muskrats. Raccoons and large owls are also important enemies, and snapping turtles, muskellunge, gars, bowfin, and bass eat a few muskrats. Foxes, cats, and dogs also destroy some.

Muskrats are pests in a farm pond. By tunneling into the dam they can cause a pond to drain; sometimes such a leak is difficult to plug.

Were it not for the name, muskrats would probably be rather popular in the stew pot. They are clean animals, and their flesh is highly palatable. In other states they have been marketed as food under the name "marsh rabbit," but such ventures have not been very successful. In the 5 trapping seasons 1968–69 through 1972–73 fur-buyers in Kentucky purchased 231,023 muskrat pelts—an average of 46,204 per year but varying from 26,995 in the 1970–71 season to 61,645 in 1968–69. Average price per pelt was $1.21; the price gradually rose from a low of 83¢ in 1968–69 to $2.07 in 1972–73.

Southern Bog Lemming;
Lemming Mouse   PLATE 24

Synaptomys cooperi (Baird)

Recognition: Total length 94–154 mm; tail 13–26 mm; hind foot 16–24 mm; weight 21–50 g. A vole with a very short tail and a shallow groove on each of the upper incisors. The pelage is rather long and silky; upperparts brown to gray, sides and underparts silvery gray.

Variation: There are two subspecies in Kentucky. *S. c. stonei* Rhoads occurs in the eastern mountain counties. A smaller race with browner pelage, *S. c. kentucki* Barbour, is found throughout the Bluegrass.

Confusing Species: This is the only mouse in Kentucky with a tail as short as the hind foot and with grooved upper incisors. In general appearance lemming mice closely resemble our 3 species of *Microtus*.

Kentucky Distribution: Locally distributed throughout eastern Kentucky and the Bluegrass, and westward along the Ohio River to Livingston County; fairly common.

Life History: Lemming mice are unpredictable creatures. They occur in little colonies, but large areas of apparently suitable habitat often contain only a single, small local colony. We can see no reason for such a distribution.

The favorite habitat in central Kentucky is a dense stand of bluegrass, especially one containing a little brush, a fallen tree, rocks, occasional shrubs, or other diversive features. Here the little animals make surface runways similar to those of the prairie vole but not as distinctive; apparently they do more foraging away from runways than *Microtus* species do.

Bog lemmings often occupy small bits of habitat that are not large enough to support meadow voles or prairie voles. Thus a roadside brush pile liberally laced with bluegrass or other herbaceous vegetation, or a fallen tree in a heavily grazed pasture, may provide a home for lemming mice. A small glade in the forest, a sphagnum bog, or a rock pile may also harbor one or more of these little creatures. In eastern Kentucky, hillsides with a good cover of broomsedge also sometimes support this species.

With their runways, little piles of cut grass, and occasional holes in the ground, bog lemmings produce signs similar to those of the meadow vole and prairie vole. One sign, however, is distinctive: droppings of the bog lemming are bright, shiny green, and once you are familiar with them they are unmistakable. *Microtus* droppings vary from dark green to black.

The nest is usually placed in a shallow burrow, but occasionally it is on the surface, well concealed in a tangle of grass or under a piece of trash. The nest is a loose structure, 100 to 200 mm in diameter, consisting of dry grass, with a lining of fine grass, fur, or feathers. It usually has 2 or 3 entrances.

Little is known about reproduction of the bog lemming in Kentucky, but we have taken pregnant females as early as January 10. Records from other states suggest a breeding season lasting into November. Litter size is usually 2 to 4; there are records of 5.

Preliminary studies of the movements and home range of the southern bog lemming indicate that in summer it may

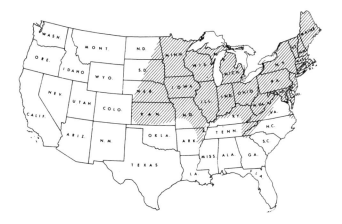

Distribution of *Synaptomys cooperi* in the United States

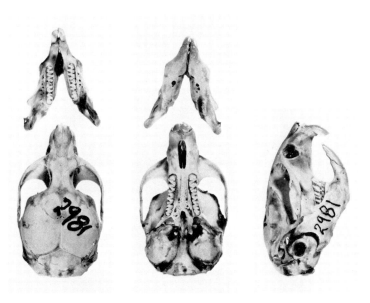

Skull of *Synaptomys cooperi*, x 2

be active at any hour of the day or night but is most active during the early morning and early evening. There is a marked decrease in activity in the hottest part of the day.

One animal ranged over 248 m² of bluegrass sod but limited most of its activity to less than 3 m from the nest. When the field was mowed the animal moved immediately and relocated in a rather sparsely vegetated but unmowed area some 15 m away, across a dry streambed. Here it ranged over 70 m² during the 13 days it resided there. A second move was to a dense stand of tall fescue beside a country road. Here it established an elongate home range, which covered only 26 m² during the remaining 30 days of the study.

A second animal inhabited 2 separate areas, which were joined by a single runway that traversed a small swampy patch covered by a tangle of Japanese honeysuckle. One of the areas measured 330 m², the other 220 m²; and the closest distance between them was 13 m.

The major part of the diet of bog lemmings consists of grass and other green leaves. The fungus *Endogone* (a favorite with several of our mice) is eaten, and raspberry seeds have been found in the stomach.

Lemming mice seem to refuse all bait and therefore are sometimes difficult to catch. This varies with habitat. In the dense bluegrass, where they use well-worn runs, they are often easily taken in traps placed across a run; but in some places

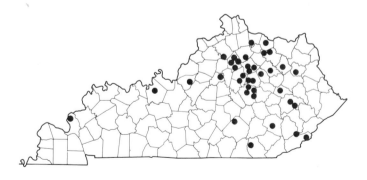

Locality records of *Synaptomys cooperi* in Kentucky

in eastern Kentucky, where a bog lemming may be the only mammal present in a bit of the terrain (as indicated by cuttings, green droppings, and absence of distinct runways) the animal may be very difficult to trap.

Lemming mice are gentle animals and adapt well to captivity. Captive mice will eat a variety of vegetable matter.

This species does little if any economic damage. Populations, which occasionally get as high as 86 per hectare, are nearly always lower than those of some of our other microtine rodents, which do cause damage at times.

FAMILY MURIDAE Old World Rats and Mice

This family has been introduced into the New World. Its members are separable from our native rats and mice on the basis of dentition; the molars are always tuberculate, with tubercles in 3 longitudinal rows. Represented in Kentucky by 2 species in 2 genera.

KEY TO THE GENERA AND SPECIES OF KENTUCKY MURIDAE

1. a. Size small, rarely exceeding 210 mm in total length; tail scant-haired: *Mus musculus*, House Mouse, p. 224
   b. Size larger, adults greater than 315 mm in total length; tail scaly: *Rattus norvegicus*, Norway Rat, p. 221

Norway Rat; Sewer Rat;
Barn Rat; Mine Rat   PLATE 24
Rattus norvegicus (Berkenhout)

Recognition: Total length 316–460 mm; tail 122–215 mm; hind foot 30–45 mm; weight 195–485 g. A large, brown rat with a long, scaly tail scantily covered with short hair. Upperparts brown to grayish; slightly darker in the middle of the back. Underparts, including the feet, nearly white.

Variation: No geographic variation is recognized. This species includes the common laboratory rat, of which various strains and mutant color forms are bred.

Confusing Species: The rice rat, *Oryzomys palustris*, is somewhat similar to a young Norway rat but is more slender and has a thinner tail, especially at the base; furthermore, the skulls and teeth of the 2 species are distinctive. The woodrat, *Neotoma floridana*, is easily separated by its well-furred, bicolored tail, its gray color, and the sharp line of separation of the darker dorsal color from the white of the underparts.

Kentucky Distribution: Statewide, wherever man has settled.

Life History: This is the worst of all mammalian pests of mankind. A native of Japan, it was brought to Europe sometime before the year 1553 and first appeared in North America about 1775. Living aboard ships, rats were distributed around the world and now occupy most places on earth that are inhabited by man.

In Kentucky Norway rats can be found in habitats ranging from the sewers and tenements of the largest cities to the gardens of the suburbs and the outbuildings, animal quarters, and open fields of the farmland. Wherever man provides suitable food and cover, Norway rats will soon appear; they reach their greatest abundance in open garbage dumps.

The rat is among the most omnivorous of animals. Anything that man considers food is food for rats. Garbage, grain stores, and spillage from the feedlot are favored. Hungry rats will attack and eat young animals and may even inflict bites on human infants.

Rats are the agents of several diseases transmissible to man. Bubonic plague and murine typhus, 2 of the most important

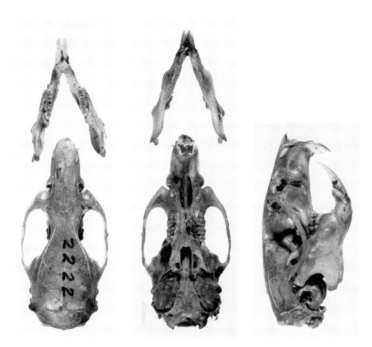

Skull of *Rattus norvegicus*, x 1.3

diseases in human history, are carried by rats and transmitted to man by the bite of rat fleas. Tularemia, trichinosis, the bacteria that cause food poisoning, and rat bite fever are also carried by these rodents.

Rats are primarily nocturnal but are occasionally seen about in the daytime. One soon becomes aware of their presence. They are active diggers, making many burrows, about 75 mm in diameter, with a conspicuous pile of dirt leading from the entrance. Gnawed holes in outbuildings, especially those containing corn or other grain, are another common sign of rats.

As evening approaches, rats become active and can be seen and heard scampering noisily about. Squealing and fighting are common, especially when rats are numerous. Although the home range is only about 0.1 hectare, an area of this size may occasionally harbor 200 rats.

Rats occasionally take up residence outside the authors'

homes in south Lexington. In winter they are attracted by scraps from the bird feeders, and in summer they invade the garden for corn, beans, peas, and tomatoes. Their burrows betray their presence and identity but we are sometimes fooled by an ambitious chipmunk that piles dirt outside his door.

Rat traps are usually effective in capturing these pests, but occasionally they become wary. Warfarin is an effective poison for rat control. However, there are recent reports of rats' developing immunity to this agent.

If a rat is captured in a steel trap or a live trap, take care in handling it. This is one of the most vicious and aggressive of our mammals.

Rats breed throughout the year. The gestation period is 21 or 22 days, and the litter size ranges from 2 to 14 but usually is 6 to 8. The young are born in a nest made of any available soft material and placed on the ground under almost any kind of shelter: a pile of trash, a woodpile, a pile of hay or manure, or a building. Only occasionally is the nest made in a burrow.

The young are blind, naked, and helpless at birth. Their eyes open at 14 to 17 days of age, and the young are weaned at about 3 weeks. By the time the young are 3 or 4 months old they are producing litters of their own. There is a record of an 8-week-old female producing and raising a litter of 11. Six to 8 litters are produced in a year.

Rats are excellent fighters and able to hold their own against some would-be predators. Many dogs and cats fear the full-grown rat. However, some cats learn to take them with ease and can be effective at keeping the population down. Hawks, owls, weasels, and other native predators occasionally eat rats.

House Mouse   PLATE 25

Mus musculus Linnaeus

Recognition: Total length 130–198 mm; tail 63–102 mm; hind foot 14–21 mm; ear 11–18 mm; weight 18–30 g. A small, gray

House mouse, *Mus musculus*. The shredded paper and black droppings are signs of house mouse activity.

mouse with a long, scaly, nearly naked tail. The nose is rather pointed. The fur generally has a somewhat greasy texture, even in animals taken outdoors.

Variation: No geographic variation is recognized in Kentucky. Many subspecies have been described around the world, but transportation and introduction by man has led to taxonomic chaos in the group. Individual variation in color ranges from yellowish to nearly black above; the underparts vary from gray to white. This is the species from which laboratory and pet-trade mice are derived. Most such mice are white (albino), but a variety of mutant colors and patterns occurs.

Confusing Species: Most similar to the harvest mouse, *Reithrodontomys humulis*, which can be distinguished by its grooved upper incisors. *Peromyscus* species lack the pointed nose and greasy fur texture; they also have larger eyes and a sharp line of demarcation between the colors of the back and belly.

Kentucky Distribution: Statewide, wherever man resides. Also resident in weedfields and brushland away from man-made structures. Apparently absent from deep woods.

Life History: The little house mouse, accidentally introduced from Europe via the ships of the earliest settlers, is one of

 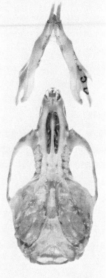 

Skull of *Mus musculus,* x 2.3

mankind's most obnoxious pests. Although not as large, vicious, or destructive as the Norway rat, the house mouse is more abundant and widespread.

The house mouse occupies buildings of all sorts—even those containing a minimum of food. However, barns and granaries are its preferred habitat, and remarkably high populations sometimes develop therein. In a single room of a seedhouse in California 235 house mice were trapped within a year, and during a mouse outbreak in the central farmlands of California the population was estimated at 200,000 per hectare. We have trapped in farm buildings where a dozen or more traps would each catch a mouse and where mice could be found by lifting a stock feeder, hay bale, or nearly any other large object on the floor.

House mice are chiefly nocturnal. They are rather secretive animals, often first seen by the householder when they appear in traps set for them. They announce their presence by nocturnal scampering and gnawing and, most disagreeably, by the characteristic little black pellets left where they have been foraging on the pantry shelf.

Almost anything that man considers food is also food to the house mouse. Flour, grain, and other stored foodstuffs are raided by these pests, which do not hesitate to chew through a cloth or wooden container to get at the supply. Raw or cooked, animal or vegetable, quality or garbage—all food seems acceptable to the house mouse.

Weed seeds and insects can also sustain the species; so it has taken up residence throughout Kentucky in places where it need not be dependent upon man. Weedfields, marshland, and grass-lined streams are the favored habitats of these feral house mice, and a line of traps in such places will often capture one or more.

Occasionally, feral house mice become so abundant as to suggest that the native species would become entirely displaced. Sometimes nearly every trap will hold a house mouse. However, this is a temporary phenomenon; the house mouse population soon falls, and the native species regain their numbers. For all their ups and downs, house mice have adapted quite well to the Kentucky countryside and are firm members of our wild fauna of small mammals.

The feral house mice take advantage of whatever is available to them. They do not make distinct trails of their own, but they readily use those of other animals. They do some digging but seem to prefer holes and shelters made by other animals.

Breeding occurs throughout the year. After a gestation period of 18 or 19 days the young are born in a nest, which is a loose structure of rags, paper, grass, or other soft material. In man's environs the nest is well hidden in the walls of a building, in a pile of trash, in a cupboard drawer or the like. We do not know where feral mice place their nests when no structure or trash pile is available.

Litter size averages 6 but ranges from 3 to 10. The young are born naked, blind, and helpless. They are furred at 10 days of age, their eyes open at 14 days, and they are weaned at 3 weeks. The young begin breeding at 7 or 8 weeks of age, and as many as 13 litters may be produced by a female in a year.

Feral house mice fall prey to many species of birds and mammals. Hawks, owls, shrikes, weasels, and cats feed upon them, in some cases making house mice an important part of the diet. In towns the main or sole predator is the house cat; on the farm various snakes, raccoons, and other predators often catch house mice in the outbuildings.

Remarks: House mice are easily taken in traps, and the population can often be controlled this way. Poisons are also effective. Houses and granaries are usually built today to exclude mice (and rats). Where livestock is fed, however, no effective way has been developed to exclude these pests.

Murine typhus and several other diseases of man have been found in the house mouse; however, it is not nearly as important in this regard as the Norway rat.

FAMILY ZAPODIDAE   Jumping Mice

This family is characterized by enlarged hind legs, which are admirably suited for jumping, and an extremely long tail. Jumping mice do not make runways, as do many of our small mammals; and, because they are quite secretive, they are rarely encountered. The family is represented in Kentucky by 2 species in 2 genera.

KEY TO THE GENERA AND SPECIES OF KENTUCKY ZAPODIDAE

1. a. Tail tip usually black; upper cheek teeth 4 on each side:
       Zapus hudsonius, Meadow Jumping Mouse, p. 229
   b. Tail tip usually white; upper cheek teeth 3 on each side:
       Napaeozapus insignis, Woodland Jumping Mouse, p. 233

Meadow Jumping Mouse PLATE 25

Zapus hudsonius (Zimmerman)

Recognition: Total length 188–250 mm; tail 112–144 mm; hind foot 28–32 mm; weight 15–20 g. A pretty little mouse with a very long tail and large hind feet. A broad, dark band runs the length of the yellowish-brown back; underparts white. The upper incisors are grooved.

Variation: Too few specimens are available to study geographic variation of jumping mice in Kentucky; however, studies on specimens from neighboring states indicate that 2 subspecies are present. The small, brightly colored *Z. h. americanus* (Barton) occurs in most of Kentucky. Specimens from the western part of the state have been referred to a larger, paler race, *Z. h. intermedius* Krutzsch.

Confusing Species: Only the woodland jumping mouse, *Napaeozapus insignis*, is similar. It can be distinguished by its much brighter color, the white pencil of fur on the tip of the tail, and the 4 upper cheek teeth on each side.

Kentucky Distribution: Western and Central Kentucky. Apparently absent in the Inner Bluegrass and the Cumberland

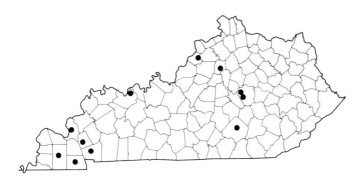

Locality records of *Zapus hudsonius* in Kentucky

Plateau where intensive trapping over many years has failed to reveal its presence.

Life History: This is a northern species—one of the small mammals characteristic of the grassland marshes, lakeside weeds, and open, weedy, wet woods of the Far North. Kentucky is near the southern edge of the range.

The preferred habitat of the jumping mouse is grassland, but dense grass is not satisfactory. Tall, weedy grassland, where the plants have enough space between them for the mice to forage, is good jumping mouse habitat; such areas are usually in moist lowlands, frequently next to a stream.

Zapus leaves little sign; the collector must rely on intensive trapping efforts in a likely-looking weed patch. In the northern states, where these mice are common, piles of glumes from the grass heads upon which they feed sometimes indicate their presence. Sometimes they climb the thickly-standing stalks to cut the heads, but more often they cut a stalk into pieces 5 to 10 cm in length until the grass head slides down within reach.

Jumping mice are nocturnal creatures, only rarely active by day. Instead of using runways, they forage by creeping about over the ground. Food consists of the seeds of grasses, the fleshy fruits of various plants, the fungus *Endogone*, and various insects.

Bait is seldom attractive to these mice. When caught in traps, they are as likely to have stepped on the treadle with a large hind foot or sprung it with the long tail as to have been taken with the head to the bait. They apparently like to explore a small open space, and a favored way of catching them is to mash down a few weeds and scatter traps in the cleared area.

Occasionally a person walking through a weedfield or mowing a meadow disturbs a jumping mouse. The animal flees with a series of jumps. The length of the jumps is a matter of dispute. H. H. T. Jackson measured a leap of 20.3 dm, and other writers have reported leaps as long as 30 to 35 dm.

Distribution of *Zapus hudsonius* in the contiguous United States

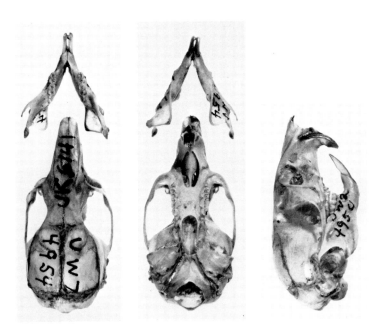

Skull of *Zapus hudsonius*, x 2.3

However, the usual leap when the mouse is first frightened is 6 to 12 dm; the next few jumps are about 3 dm each, and then the animal sits quietly. If the observer has not carefully noted where the animal was last seen, he is likely to have lost it, because a jumping mouse sitting motionless on the ground is difficult to see. If the observer is careful in pursuit, however, he will find that a jumping mouse is rather easy to catch by hand.

The summer nest is made of dried grass or leaves and placed in a hummock of grass, under an overhanging clump of grass, in a rotten log or stump, or in a shallow burrow. It is about 15 cm in diameter and has a small entrance on the side.

Two litters are produced per year, the first in May or June and the second in July or August. Litter size ranges from 3 to 8 but averages about 5. The young are naked and helpless at birth, with closed ears and eyes. Hair appears about the ninth day, the ears open about the 19th day, and the eyes open between the 22nd and 25th days. At the end of the fourth week the young are weaned.

The home range of the jumping mouse is large. At one locality in Minnesota the home range averaged about 0.6 hectare, and at another locality the average was 1.1 hectares for males and 0.65 hectare for females. Habitat, season, and population density influence the size of the home range.

In the fall jumping mice become fat in preparation for hibernation. Sometime in November they retire to a nest 0.6 to 1 m below ground in a burrow made by a groundhog or other animal; there they hibernate until April. The mouse curls into a ball, with its feet and nose close together and its long tail curled like a watchspring.

Jumping mice do not hesitate to take to the water; they are accomplished swimmers and divers, using the large hind feet for propulsion.

In northern regions where jumping mice are common they are fed upon by a large variety of predators. Owls, mink, and weasels seem to be their most important enemies, but foxes, snakes, frogs, and fish eat a few. Hawks capture them rarely.

Jumping mice are of little economic significance even where common; they are seldom encountered by man and do not damage crops.

Woodland Jumping Mouse PLATE 25

Napaeozapus insignis (Miller)

Recognition: Total length 204–259 mm; tail 115–160 mm; hind foot 28–34 mm; ear 15–19 mm; weight 16–32 g. A beautiful mouse, bright yellow to orange on the sides and about the face. A broad band of brown runs the length of the back, and the underparts are white. The tail is quite long and is bicolored: brown above and white below, with a solid white tip. The upper incisors are grooved; the hind feet are large.

Variation: No geographic variation is recognized in Kentucky. The subspecies here is *N. i. roanensis* (Preble).

Confusing Species: Only the meadow jumping mouse, *Zapus hudsonius,* among Kentucky mice has such a long tail and large hind feet. The woodland jumping mouse can be separated by its brighter color, the white tail tip, and the absence of the tiny upper premolars nearly always found in *Zapus.*

Kentucky Distribution: This is a northeastern species that ranges down the Appalachian Mountains to South Carolina and Georgia. Until recently we thought its distribution in Kentucky was limited to the woodlands about the summit of Big Black Mountain in Harlan County. However, specimens have recently been taken along woodland streams in the lowlands of Elliott and Leslie counties.

Life History: This is an animal of cool, moist woodlands that are open enough to provide good stands of herbaceous ground cover. Because such locations often border streams, woodland

Distribution of *Napaeozapus insignis* in the United States

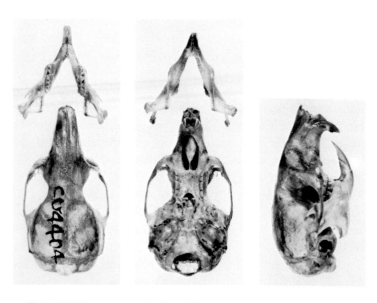

Skull of *Napaeozapus insignis*, x 2.1

streams are generally considered to be the favored habitat; however, excellent habitat, containing good populations of jumping mice, is sometimes found far from streams.

About the summit of Big Black Mountain running water is scarce except for a few small springs. The distribution of woodland jumping mice in the forest is sparse and has no apparent relationship to the springs. Streams edged with grass, sedge, and alder, favored by this species in other regions, are not available on Big Black Mountain. This mouse should be sought in eastern Kentucky where woodland springs and small streams, bordered by grasses and rank herbaceous vegetation, drain cool northern slopes.

Woodland jumping mice are mainly nocturnal but are occasionally active in the morning and early evening. They forage on the damp forest floor for the subterranean fungus *Endogone*, which makes up about a third of their diet·and is also an important food for several other small mammals.

Plant matter makes up approximately 80% of the diet and insects most of the remainder. Blueberries, raspberries, small seeds and nuts, miterwort, mayapple, fern fronds, and leaves of other vegetation are eaten. Seeds make up about one-fourth of the diet. The seeds of jewelweed, or wild touch-me-not, a plant that usually grows where these jumping mice are found, are especially important. Adult beetles and craneflies, larval insects, an occasional centipede, and other small invertebrates have been found in the stomachs of these mice.

Breeding occurs after the mice emerge from hibernation, in April or May. The gestation period is somewhere in the range of 23 to 30 days. The young are born in June. Litter size ranges from 2 to 7; the average is 4 or 5. Newborn young are naked, blind, and helpless, and they weigh about 0.9 g each. They grow slowly, changing little during the first week. The ears unfold on the 10th day, but the auditory canal remains closed. By the 12th day fine hairs are visible, and at age 24 days the young are well furred; however, their eyes and the external auditory canal do not open until the 26th day. Young first appear above ground in late June and are weaned when about

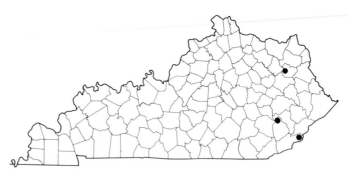

Locality records of *Napaeozapus insignis* in Kentucky

34 days old. A second litter is born in August or September.

The nest, of dry leaves, is usually located in a burrow but is often made in a natural cavity under a rock or log. Usually the burrow of another animal is used, but these jumping mice are reported to dig their own burrows occasionally.

When startled, woodland jumping mice can leap as much as 18 dm in a single bound. They then proceed in leaps 5 to 10 dm long and half as high. After several hops the animal abruptly stops and takes cover, remaining motionless unless disturbed again.

This animal climbs well in bushes but does not ascend trees. It is a good swimmer both on the surface and underwater but has poor endurance, becoming exhausted in about 3 minutes.

As the days shorten in October, woodland jumping mice fatten for hibernation. Within a period of 2 weeks they deposit fat to the extent of as much as one-third of the body weight. Because they do not store food, this fat must provide their winter nourishment. Sometime in late October or in November our jumping mice retreat to an underground nest below the frostline to hibernate until April.

Several studies of the home range and population density of woodland jumping mice have been published. Populations ranging from 0.64 to 59 mice per hectare have been reported. Home ranges have been estimated at 0.4 to 2.6 hectares for females and 0.4 to 3.6 hectares for males.

These mice of the wildlands rarely come into contact with man; even in regions where they are most abundant they do not impinge upon man's interests. In Kentucky this most beautiful mouse is extremely rare, and on aesthetic grounds we consider it one of our state's most valuable treasures.

ORDER CARNIVORA *Carnivores*

Members of this order show the greatest diversity in size seen in any order of Kentucky mammals: from the least weasel, which is smaller than a chipmunk, to the brown bear. The order is characterized by teeth adapted to cutting or tearing flesh. The canines are long and pointed. Most carnivores possess specialized cutting teeth in the upper and lower jaws— the carnassial teeth. As the names implies, most carnivores are meat-eaters, but many of them—notably the skunk, raccoon, and fox—also eat considerable quantities of vegetable matter. Many of the carnivores are of considerable importance in the fur trade. Essentially worldwide ,in distribution, the order has about 285 species in 10 families. Kentucky has 10 genera in 5 families.

KEY TO THE FAMILIES OF KENTUCKY CARNIVORA

1. a. The larger molars with crowns of the crushing type and without a conspicuous cutting edge; feet plantigrade (sole-to-ground): 2

   b. The larger molars with crowns at least partly of the cutting type; feet mainly digitigrade (toes-to-ground): 3

1a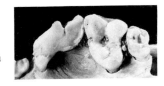

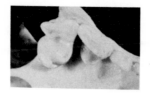 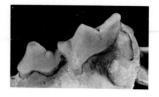

1b, 3b   3a

2. a. Tail short; teeth 42: Ursidae, bears, p. 250

   b. Tail long, ringed; teeth 40: Procyonidae, raccoons and allies, p. 254

3. a. Molars without crushing surfaces; claws completely retractile: Felidae, cats, p. 278

   b. Rear molars with evident crushing surfaces; claws not retractile or only partly so: 4

4. a. Tooth rows relatively short; teeth 38 or fewer: Mustelidae, weasels and allies, p. 259

   b. Tooth rows long; teeth 42; body doglike: Canidae, foxes and allies, p. 238

Family Canidae   Foxes and Allies

Digitigrade carnivores with blunt, nonretractile claws; forefoot with 5 toes, hind foot with 4. Muzzle usually elongate; canines large and rather blunt. Represented in Kentucky by 3 species in 3 genera.

KEY TO GENERA AND SPECIES OF KENTUCKY CANIDAE

1. a. Tail vertebrae more than half as long as the head and body; hind foot less than 175 mm; width of nose pad less than 19 mm: 2

b. Tail vertebrae less than half as long as the head and body; hind foot more than 175 mm; width of nose pad more than 19 mm: *Canis latrans*, Coyote, p. 239

2. a. Legs and feet reddish-brown; tail tipped with black; cranial ridges lyrate forming prominent ridges on the posterolateral surfaces of the skull: *Urocyon cinereoargenteus*, Gray Fox, p. 247

   b. Legs and feet blackish; tail tipped with white; cranial ridges nearly meeting, and forming a prominent sagittal crest posteriorly: *Vulpes vulpes*, Red Fox, p. 243

COYOTE OR DOG?

In view of the appearance of the coyote within the state in recent times and its apparently increasing numbers, the following key is presented.

1. a. Width of rostrum less than 18% of the greatest length of the skull; upper incisors always close-set and even: *Canis latrans*, Coyote, p. 239

   b. Width of rostrum more than 18% of greatest length of skull; upper incisors usually with spaces between them: *Canis familiaris*, Domestic Dog

Coyote; Brush Wolf;
Prairie Wolf   PLATE 26

Canis latrans Say

Recognition: Total length 1,052–1,320 mm; tail 300–394 mm; hind foot 177–220 mm; weight 9–16 kg. A large, gray animal similar in general appearance to a small police dog. Color gray to tawny above; dorsal hairs tipped with black. Throat and belly white. Ears, feet, and outer sides of hind legs washed with brown.

Variation: The taxonomic status of coyotes in the East is confused, because various subspecies have been introduced by man, the species has also spread eastward naturally, and coyotes have hybridized with dogs and wolves. The result is a highly variable population. In New York and New England coyotes vary in color from light gray to black, and many individuals are larger than any western coyote. For lack of specimens taken in Kentucky, we are unable to assess variation here.

Confusing Species: The gray fox, *Urocyon cinereoargenteus*, is smaller, has a black stripe down the top of the tail, and has extensive brown areas on the tail, belly, legs, and neck; also, its skull is distinctive. Some domestic dogs are difficult to distinguish from coyotes (see special key).

Kentucky Distribution: The only specimen we have seen is from Clark County, in the Inner Bluegrass. Possibly the animal was an escaped captive or was released by foxhunters.

The natural range of the coyote has been expanding eastward in the southern states. A specimen in the U.S. National Museum was taken in 1951 in Henry County, Tennessee, next to Calloway County, Kentucky. This is reason to suspect that the coyote occurs naturally in the Purchase.

Life History: The coyote of the Old West is holding its own in spite of intensive persecution by man. Although traps and poisons have taken a heavy toll and have nearly eliminated coyotes in parts of the rural West, the wily animals have invaded man's own precincts, even to the extent of becoming established in the suburbs of Los Angeles and other cities.

An aid to range extension has been the clearing of forests in the East. Here, too, the war on predators is less intensive and the coyote is more likely to be accepted in peaceful coexistence.

In the East coyotes mainly inhabit open brushlands. Woodland borders and the brushy growth that invades the hardwood forest after cutting or burning also provide shelter for coyotes.

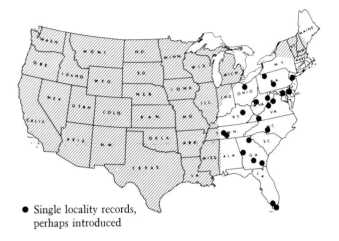

● Single locality records, perhaps introduced

Distribution of *Canis latrans* in the contiguous United States

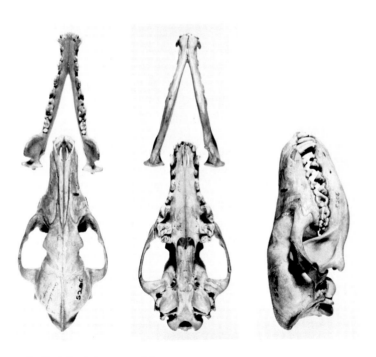

Skull of *Canis latrans*, x 0.2

Marshlands and fields interspersed with thickets are also used. The characteristic yapping and howling of the coyotes—the distinctive and beautiful night-music of the West—is rarely heard in the East. And the eastern coyotes are seldom seen; apparently they are much more shy than their western brothers, which, although largely nocturnal, are occasionally seen in the daytime.

The den usually is a burrow but sometimes is a crevice in a cliff or among boulders. Occasionally the coyote digs its own burrow, but more often it enlarges a burrow made by a woodchuck, skunk, or fox. The entrance, about 3 dm in diameter, is usually hidden by vegetation.

The burrow is rather shallow—usually not more than 10 or 12 dm deep. It may extend 6 m or more and have 2 or 3 entrances. The young are born in a chamber nearly a meter in diameter. No nesting material is used.

Breeding occurs in February. The young are born in April, after a gestation period of 60 to 63 days. Litter size averages 5, commonly ranges from 4 to 10, and occasionally goes as high as 19. The pups are born blind and lightly covered with short, rough, yellowish-brown fur. Their eyes open in about 10 days. When about a month old the fuzzy little whelps begin to play outside the den, and at 8 or 9 weeks of age they are weaned.

The food of the coyote averages 98% animal matter. Vegetable matter consists mostly of wild fruit, which makes up as much as 6% of the diet in late summer and fall. Mammals make up 96% of the animal matter; birds, 3%; reptiles, amphibians, and fish, 0.1%; and invertebrates, (mostly insects), about 1%. Sheep and deer form a significant part of the diet. Most of this is carrion, but lambs and fawns are often taken when the growing whelps are being fed. Both the mother and the father coyote bring meat to the den at that season.

Coyotes are host to a large and varied assortment of parasites and diseases. These cause many deaths; but man is the major enemy. Coyotes are extensively shot, poisoned, and trapped, and some are killed on highways.

There is intense controversy about the economic status of the coyote in the West. Although the coyote is an important factor in keeping down the population of small mammals that compete with sheep and cattle for available forage, coyote predation sometimes is extensive in the lambing season, and occasionally calves are attacked. Probably to the cattleman the good outweighs the damage, but not so for the sheepman. In Kentucky coyotes are too scarce to be of any economic significance.

Red Fox PLATE 26

Vulpes vulpes (Linnaeus)

Recognition: Total length 900–1,136 mm; tail 330–406 mm; hind foot 145–175 mm; ear 88–110 mm; weight 2.7–5.8 kg (males larger than females). A beautiful reddish-yellow animal with the general appearance of a small dog. The nose pad, the back of the ears, the legs, and the feet are black; the cheeks, throat, and belly are white. The long, heavily furred tail is tipped with white. Footprints are like those of a small dog but narrower.

Variation: Various color phases occur but most are very scarce in Kentucky. These include the so-called silver fox (a melanistic mutant, in which a black pelage is frosted with white), and the cross fox, in which the pelt is mixed with gray and yellow. No geographic variation is recognized in Kentucky. The subspecies here is V. v. fulva (Desmarest).

Confusing Species: The slender red fox is not likely to be confused with any other animal.

Kentucky Distribution: Statewide; common.

Life History: The red fox is a characteristic animal of the farmlands and other open areas of Kentucky. It is principally

Distribution of *Vulpes vulpes* in the contiguous United States

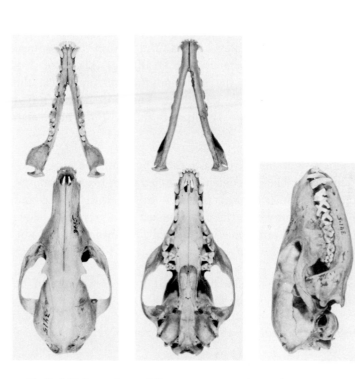

Skull of *Vulpes vulpes*, x 0.3

nocturnal and thus seldom encountered by man, but its narrow, doglike footprints can commonly be seen on dusty roads or in snow. Occasionally one sees a fox foraging for meadow mice by day; more often it is jumped from its day bed in tall grass. The fox den is usually located just inside a piece of woodland bordering on a field. Streambanks and spaces beneath abandoned buildings are also favored sites, and natural rock crevices are sometimes used. The den is used for raising the pups and as a bedding site in cold weather. Most dens are modified burrows of woodchucks, but sometimes a fox digs its own. A burrow may be used for several years. The strong, characteristic odor of an occupied fox burrow betrays its inhabitants. Burrows are usually 4 to 6 m long and about a meter deep, and there may be one entrance or several. The entrance is about 30 cm in diameter. Breeding dens usually have branched tunnels.

Breeding occurs in January or February. After a gestation period of 53 days a litter of 2 to 10 (average 5) is born in March or April; there is but one litter per year. At birth the young are lightly covered with fine, dull-gray hair, but the distinctive white tip on the tail is evident. The eyes open at 9 days, and within 3 weeks the pups are playing outside the den. Both parents guard the young and bring food. The pups are weaned at age 2 months and may breed the following winter.

The home range is usually within a radius of 1.5 km from the den. The fox follows trails, paths, and roads in its foraging, and its presence is betrayed to the careful observer by the small, doglike droppings, consisting largely of fur, deposited on the trail.

Although the fox can attain a speed of 48 km per hour, it does not chase its prey for any appreciable distance. Instead, it walks slowly along a trail or the edge of a field, investigating each fallen log, brush pile, or burrow for prey. If a log, rock, or other elevation is encountered, the fox will get upon it to survey the surrounding area.

When the young are being raised, foxes often hunt in pairs. Approaching a brush pile, the foxes separate to go around it:

Red fox tracks, x 0.5

a rabbit or mouse jumped therefrom is then easier prey. At a culvert one fox will go through it while the other waits at the other end to capture any animal that may have been resting within.

Cottontail rabbits are the most important food of the fox, making up nearly half the diet. Meadow voles, deer mice, and other small mammals account for most of the rest. Larger animals, such as rats, woodchucks, muskrats, skunks, weasels, and opossums, are also eaten. Fruit, grass, corn, insects, birds, fish, amphibians, and reptiles round out the varied diet of the red fox. The animal is an opportunist, feeding mostly upon what is most readily available. Only rarely does one take to feeding upon domestic chickens.

Unlike the coyote, the fox is seldom noisy. Occasionally it utters a few short, high-pitched yaps and barks, sometimes ending in a shriek.

The economic status of the fox has long been a matter of dispute. Its effects on game birds and domestic fowl are negligible. Hunters decry its toll of cottontails, but good cover is more important to rabbit populations than is predator pressure. Foxes exert a beneficial effect in keeping control of the population of other mammals; on balance, they are probably beneficial. To the foxhunter, who chases them with hounds, and to the naturalist, who delights in the sight of one, red foxes are a valuable part of our natural fauna.

In the 5 trapping seasons 1968–69 through 1972–73 fur-buyers in Kentucky purchased 4,439 red fox pelts; the average was 887 per year, from a low of 636 in 1970–71 to highs of 1,173 in 1968–69 and 1,113 in 1972–73. Average price per pelt was $6.34; the price advanced from $4.22 in 1969–70 to $11.98 in 1972–73.

Gray Fox

PLATE 27

Urocyon cinereoargenteus (Schreber)

Recognition: Total length 800–1,125 mm; tail 275–443 mm; hind foot 100–150 mm; weight 3.2–5.4 kg. Upperparts grizzled gray, with long guard hairs tipped with white and black. Throat white; chest, flanks, underside of the tail, and inner side of the hind legs tawny. The skull has a widely separated pair of ridges on the braincase.

Variation: Males average about 10% larger than females. Color mutants are rare, but black individuals have been reported. No geographic variation is recognized in Kentucky. The subspecies here is *U. c. cinereoargenteus* (Schreber).

Confusing Species: The gray fox cannot be confused with any other Kentucky mammal. The skull is also distinctive; the prominent lyrate ridges on the braincase distinguish it.

Kentucky Distribution: Statewide; more common in forested regions than in open areas.

Life History: The gray fox is a typical animal of the hardwood forests of eastern Kentucky, the wooded or brushy hilly terrain of most of the state, and the heavily wooded bottomlands of western Kentucky.

Round, evenly spaced footprints running in a nearly straight line in the mud or snow on a forest trail are the sign of the gray fox. The prints are similar to those of a large house cat, except that the claws show. The fur-filled droppings are scarcely distinguishable from those of the red fox, but they tend to be heavier and more creased into individual pellets.

Gray foxes are chiefly nocturnal, spending the day asleep in a hollow log, rock pile, or other den or on a sunny hillside. Occasionally, however, they are abroad and foraging by day. Like weasels and some other carnivores, they will respond to a man-made squeak in imitation of an injured mouse or bird.

Distribution of *Urocyon cinereoargenteus* in the United States

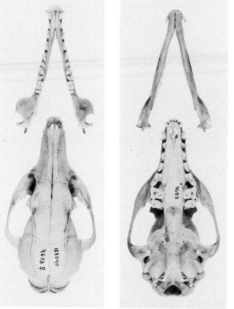

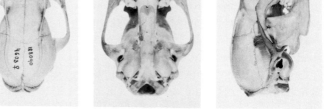

Skull of *Urocyon cinereoargenteus*, x 0.4

On several occasions we have had a gray fox rush at us from over a rise in the road or from a trail out of the brush, only to stop and stand in astonishment upon learning the source of the squeak.

This is our only member of the dog family that can climb trees. Not only can a gray fox go up the trunk of a tree almost like a cat; it will jump from branch to branch. When pursued by dogs, this fox quickly trees or goes into a den; thus it has little interest for the foxhunter.

The home range of the gray fox is smaller than that of the red fox. It is seldom more than 1.5 km in diameter and often half that size.

Mating occurs in February or March. After a gestation period of 51 to 63 days the single litter of 3 to 5 blind and nearly naked pups is born. The den is in a hollow tree or log, a rock crevice, or a burrow in the woods. The gray fox rarely dens in an open field, as the red fox sometimes does.

The pups are born in a scant nest of leaves, grass, and fur. Their eyes open at 9 days of age. When the pups are 2 or 3 weeks old both parents begin to bring home food for the family. At about 3 months of age the young begin to hunt for themselves, but the family tends to remain together until fall, when the pups are nearly full-grown.

The food of the gray fox is similar to that of the red fox. Cottontail rabbits are the main food item, but meadow voles, white-footed mice, and other rodents are also important. Birds, insects, and fruit make up most of the rest of the diet.

The economic significance of the gray fox in Kentucky is a matter of dispute. Many sportsmen consider its feeding upon rabbits to be detrimental. It occasionally takes quail, grouse, and chickens. On the plus side is the gray fox's effect on the rodent population.

In the 5 trapping seasons 1968–69 through 1972–73 fur-buyers in Kentucky purchased 8,852 gray fox pelts—an average of 1,770 per year. Purchases varied from 1,297 in 1971–72 to 2,798 in 1972–73. Average price per pelt was $3.22; prices ranged from $2.05 in 1970–71 to $6.34 in 1972–73.

FAMILY URSIDAE Bears

Members of this family are large, plantigrade carnivores with a rudimentary tail. There are 42 teeth; the molars are large and have crowns of the crushing type. The family is represented in Kentucky by a single species.

Black Bear PLATE 27

Ursus americanus Pallas

Recognition: Total length 1,270–1,780 mm; tail 80–125 mm; hind foot 215–280 mm; weight 101–225 kg. A large, robust, short-tailed, black (occasionally brown) animal.

Variation: Males are about 10% larger than females. Most individuals are black, but there are occasional variants toward brownish and a few are cinnamon brown. No geographic variation is recognized in Kentucky. Our subspecies is *U. a. americanus* Pallas.

Confusing Species: No other Kentucky mammal could be confused with the black bear, whose appearance is distinctive. Furthermore, it is our largest native animal.

Kentucky Distribution: Formerly statewide; now may be extirpated. Bears are occasionally reported from the counties bordering on Virginia or Tennessee, especially Harlan and Bell, and there have been reports from Powell and Fleming counties within the past 30 years. Whether or not bears actually occur in Kentucky today and, if so, whether they are resident or are wanderers from adjacent states, is not firmly established. There are no known resident populations of bears in any of the counties of Virginia or Tennessee that border upon Kentucky.

Black bear scratching himself on a tree trunk

Life History: The bear is an animal of the forested wilderness and retreats with the advance of mankind except in special situations, such as in Great Smoky Mountains National Park, where bears have successfully adapted to man's activities. In the eastern states today bears are restricted to a few wild regions: extensive forests in the north and the nearly impenetrable swamps of the Deep South, where man and dogs seldom venture. Kentucky now seems to lack habitat suitable to maintain a population.

Bears are wary, timid, and nocturnal and thus are rarely seen by men. In the Smoky Mountains, however, they visit garbage cans in the campgrounds nightly, and in some parks in the western states they show no fear of man.

Bears wander considerable distances; a normal evening's foraging may cover several kilometers. Old males commonly range 15 km, and there is evidence that some wander 130 km or more from their established home range.

As evening approaches, bears leave their daytime beds in the dense vegetation and begin to forage. Their nightly activities are evidenced by rotted logs torn up in search of insects and mice, diggings around trees and stumps, and the marks of wallowing in patches of wild berries. Large, soft droppings, sometimes consisting almost entirely of seeds of berries or other fruit, can be seen on feeding grounds and along forest trails. Footprints, somewhat like those of a barefoot boy but broader

Distribution of *Ursus americanus* in the contiguous United States

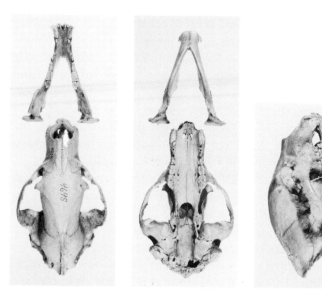

Skull of *Ursus americanus*, x 0.1

and, of course, with long claw marks, are another sign of bears. In the Monongahela National Forest in West Virginia, we have commonly seen signs of bear activity expressed as gnawings and claw marks on trees. A bear will reach as high as it can on a spruce tree along a trail and bite off a large piece of bark and wood. The scar on the tree, the tooth marks, and the bleeding of pitch are quite conspicuous. Bears in the high country of West Virginia also destroy, by chewing, the wooden signs erected by the U.S. Forest Service; at a few spots the foresters have despaired of trying to maintain signs, because of the bears.

The black bear is omnivorous, feeding on a wide variety of animal and vegetable matter. Insects, mice, chipmunks, fish, nestling birds and birds' eggs, honey, carrion, and garbage are eaten. Domestic livestock is sometimes taken—young pigs and sheep especially. Occasional individuals become habitual livestock-killers.

Vegetable matter makes up a major part of the diet. Bears are especially fond of blueberries, but strawberries, dogwood berries, serviceberries, elderberries, blackberries, raspberries, apples, and wild grapes are frequently eaten. The bear picks small berries with its lips and tongue. Grass, roots, acorns, and beechnuts round out the vegetable part of the diet.

As autumn approaches, bears get very fat in preparation for a winter of inactivity. In late November or December a bear retires to its winter den. This may be a shallow cave, a shelter among boulders, a large hollow log, a cavity beneath the roots of an overturned tree, a pile of brush, or simply the sheltered space beneath the lowest branches of a hemlock or spruce. Here the bear sleeps through the winter in a condition of somewhat lowered metabolism; but this is not true hibernation, as in some of our smaller mammals. The body temperature fluctuates between 29°C and 34°C, the respiration rate falls to 12 per minute, and oxygen consumption is only about 50 to 60% of that of summer. The bear is drowsy but is easily aroused.

Mating takes place in June and, after a gestation period of

about 225 days, 1 to 4 cubs are born in January. The cubs weigh 200 to 340 g at birth—remarkably little, considering the size of the mother. Their eyes open at about 25 days of age, by which time the youngsters are covered with short, sparse hair. By the time the mother and young leave the winter home, in March, the cubs have become quite active and playful. The mother is very attentive and will fight gallantly in their behalf. They remain with the mother for more than a year, usually sleeping with her through their first winter. The female produces a litter only in alternate years.

Bears are not sexually mature until 3 years old, and a young one is not full-grown until its sixth year.

The meat is delicious. Bears are considered prize game in those states fortunate enough to maintain a harvestable surplus.

FAMILY PROCYONIDAE   Raccoons and Allies

Members of this family are medium-sized carnivores. The tail is well developed and is prominently marked with rings. The molariform teeth are of the crushing type, and the total number of teeth is 40. Represented in Kentucky by a single species.

Raccoon   PLATE 28

Procyon lotor (Linnaeus)

Recognition: Total length 700–960 mm; tail 225–275 mm; hind foot 110–125 mm; weight 5.4–13.5 kg. A large, furry animal with a conspicuous black mask across the eyes, set off by white on the snout and above the eyes. The rather long, heavily furred tail is ringed with black and yellowish markings. Typically, the body is grizzled gray above, paler on the sides, and light gray below.

Raccoon stalking
a well-camouflaged frog

Variation: Raccoons vary from light tan to dark gray. Two subspecies occur in Kentucky. *P. l. lotor* (Linnaeus) occupies most of the state. A smaller, paler, short-haired southern race, *P. l. varius* Nelson and Goldman, occurs in the Purchase.

Confusing Species: No other mammal of regular occurrence in Kentucky could be confused with the raccoon; the combination of black mask and ringed tail is distinctive. The ringtail, *Bassariscus astutus*, also has a ringed tail but is more slender and lacks the black mask; there is only one specimen of the ringtail from Kentucky, and it may have been an escaped animal.

Kentucky Distribution: Statewide; abundant.

Life History: Raccoons prefer woodland; they especially like old hardwood stands, which contain hollow trees. However, they are also found in brushland, farmland, small towns, and other habitats.

Raccoons are strongly partial to water, foraging heavily at the edge of a stream, pond, or lake in search of crayfish, frogs, and other edibles. There is scarcely a body of water in the state where one cannot find raccoon tracks at the muddy edge.

Raccoons are nocturnal, generally emerging from the den shortly after dusk to begin their nightly prowl. In some of our

Baby raccoons, *Procyon lotor*

parks and campgrounds they appear well before dark, looking for a handout. Occasionally one is abroad in mid-day to lounge on a tree branch and sun itself.

Home for a raccoon usually is a hole in a tree, either in a large limb or the main trunk. Barns, abandoned buildings, mines, caves, and rock crevices are also used. No nest is made.

Mating occurs from January to March, and, following a gestation period of 63 to 65 days, a litter of 2 to 7 is born. The young are well formed at birth, and the blackish skin clearly shows the facial mask and tail rings. Newborn raccoons weigh about 84 g. At about 20 days of age their eyes open. The young are weaned at 10 to 12 weeks; however, they forage with the mother until well into autumn. They are ready to breed toward the end of their first winter.

The raccoon is omnivorous, and its food habits depend in large part upon what is most readily available. Generally, more plant than animal matter is taken, especially in summer. Wild cherries, grapes, plums, apples, blackberries, elderberries, hackberries, acorns, nuts, and corn are favored. Animals eaten include crayfish, clams, snails, grasshoppers, crickets, beetles, wasps, bees, moths, fish, and frogs. Turtle eggs, a few birds

Distribution of *Procyon lotor* in the United States

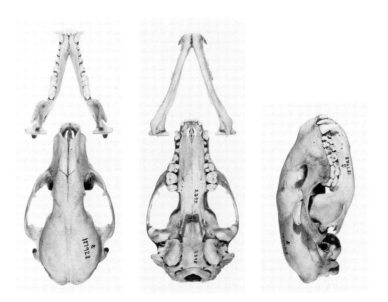

Skull of *Procyon lotor*, x 0.4

and their eggs, and an occasional small mammal round out the diet.

A raccoon gathers its food in its remarkable, handlike front paws and transfers it to the mouth. At the creek's edge it feels under stones with its paws and catches crayfish, sculpins, and darters.

Raccoons are active throughout the year. In very cold weather, however, they stay in the den for several days at a time, subsisting on the reserve of fat accumulated in late summer.

Raccoons are furbearers and are also considered game animals in Kentucky, where they are taken by trappers and are commonly hunted with dogs. To hunters and trappers the animal is a "coon." If the dogs tree a coon, it is easily taken by shooting. But if a coon takes to the water when pursued by hounds, woe be unto the dog that catches him there, for a coon can whip a dog in deep water and can sometimes drown one.

Raccoons provide man with an interesting change in table fare. The meat is delicious if it is parboiled and if the scent glands have been removed from under the legs and along the midline at the rump.

Young raccoons—playful, intelligent, interesting animals, are favored as pets. Pet raccoons often get cross, however, as they get older.

The raccoon is generally beneficial if not too numerous. It is an interesting animal, provides excellent sport for night hunting, and is of value on the fur market. Sometimes, however, it is harmful. A raccoon can really wreck the corn in your garden, and occasionally one will raid the henhouse.

In the 5 trapping seasons 1968–69 through 1972–73, furbuyers in Kentucky purchased 74,782 raccoon pelts. The average was nearly 15,000 per year, but purchases varied from a low of 9,146 in the 1970–71 season to a high of 21,355 in 1972–73. Average price per pelt was $2.40; the price range was $1.05 in 1970–71 to $4.31 in 1972–73.

FAMILY MUSTELIDAE Weasels and Allies

Members of this family are small to medium-sized carnivores, either plantigrade or digitigrade, with 5 toes on each foot. Most species have glands near the anus that secrete a vile-smelling fluid. Some are strictly carnivorous; others are rather omnivorous. The family is represented in Kentucky by 5 species in 4 genera.

KEY TO THE GENERA AND SPECIES OF
KENTUCKY MUSTELIDAE

1. a. Modified for aquatic life; toes fully webbed: *Lontra canadensis*, River Otter, p. 274
   b. Toes scarcely or not at all webbed: 2

2. a. Color black and white: 3
   b. Color not black and white: 4

3. a. White stripes variable, the posterior ones at right angles to the anterior; upper outline of the skull straight: *Spilogale putorius*, Eastern Spotted Skunk, p. 266
   b. Two white dorsal stripes variable in extent, sometimes absent; upper outline of the skull convex: *Mephitis mephitis*, Striped Skunk, p. 269

4. a. Uniformly dark brown, usually with a white spot on the throat; pads of the palm coalescent; toes of the hind feet partially webbed: *Mustela vison*, Mink, p. 262
   b. Brown above, underparts buffy or yellowish-white; tail tip black; pads of the palm not coalescent; toes of the hind foot not at all webbed: *Mustela frenata*, Long-tailed Weasel, p. 260

Long-tailed Weasel PLATE 28
Mustela frenata Lichtenstein

Recognition: Total length 285–340 mm in females, 350–431 mm in males; tail 85–123 mm in females, 115–150 mm in males; hind foot 30–40 mm in females, 40–51 mm in males; weight 85–130 g in females, 170–245 g in males. A long, slender animal with rather large, rounded ears. Brown above, yellowish below.

Variation: No geographic variation is recognized in Kentucky. Our subspecies is *M. f. noveboracensis* (Emmons).

Confusing Species: The mink, *M. vison,* is larger and darker. The least weasel, *M. nivalis,* which may occur in Kentucky, is smaller and has a short tail.

Kentucky Distribution: Statewide; fairly common.

Life History: Forest edge, brushland, fencerows, and stream-banks are the favored homes of this tireless bundle of energy. Squeaking sounds in imitation of an injured bird or mouse occasionally bring out a curious weasel from such places.

Although generally nocturnal, weasels sometimes hunt for chipmunks and meadow voles in the daylight hours. They are active summer and winter, their tracks in the snow being the best indication of their presence. The gait is a bounding gallop with the back arched, so that the hind feet are placed just behind the prints of the front feet; this makes a distinctive track of groups of 4 prints.

A weasel will climb a tree when chasing a chipmunk or other prey but does not seem at ease there. It is an inquisitive creature, emerging from hiding to investigate the slightest noise. A weasel will often stand erect to get a better view.

The den is a modified chipmunk burrow, a crevice or hole in a stone wall, a cavity beneath a stump, or some other secluded place. The nest is made of densely packed grass and

Distribution of *Mustela frenata* in the United States

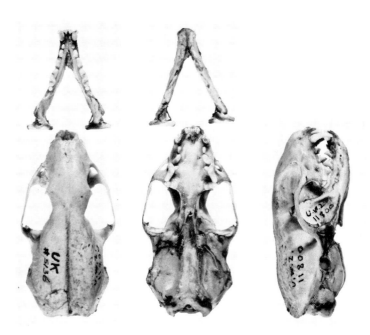

Skull of *Mustela frenata*, x 1.1

lined with mouse and shrew fur. The burrow may contain skins and bones of animals that were eaten.

Breeding occurs in July or August. The gestation period ranges from 205 to 337 days, and the fertilized ova remain undeveloped for most of this period; implantation takes place only about 27 days before birth. The litter of 6 or 8 young is born in April. The young are naked and blind at birth, and their eyes do not open until the 35th day. By 3 months of age the young are nearly mature. Young females breed in their first summer; males do not.

The weasel is an absolutely fearless hunter; it will attack a man or any other creature that tries to interfere. Its food is chiefly small mammals, especially mice of the genera *Microtus* and *Peromyscus*. Young rabbits, chipmunks, rats, shrews, moles, and an occasional bird are eaten. A weasel was once seen feeding on a garter snake.

Occasionally a weasel will invade the henhouse and kill all the young poultry it can find. There is a record of a weasel killing and feeding on baby pigs. However, because weasels are a terror to rats and mice, they are generally beneficial. Only the occasional individual is a nuisance, but it can be very destructive.

In the 5 trapping seasons 1968–69 through 1972–73 fur-buyers in Kentucky purchased 485 long-tailed weasel pelts, averaging 97 per year but varying their purchases from 27 in 1970–71 to 168 in 1969–70. Average price per pelt was about 61¢; the range was 42¢ in 1970–71 to 68¢ in 1972–73.

Mink   PLATE 29

Mustela vison Schreber

Recognition: Total length 460–575 mm in females, 580–700 mm in males; tail 150–190 mm in females, 190–230 mm in males; hind foot 60–70 mm in females, 68–80 mm in males;

weight 675–1,080 g in females, 855–1,620 g in males. A rather large animal with sleek, dark-brown fur and a well-furred tail. Color is nearly uniform, with underparts only slightly paler; there is usually a white spot beneath the chin.

Variation: Occasional individuals have white streaks or blotches on the underparts. Our subspecies over most of the state is *M. v. mink* Peale and Palisot de Beauvois. In the Cumberland Mountains a smaller, darker race, *M. v. vison* Schreber, occurs.

Confusing Species: The mink can be easily distinguished from its associates the muskrat and beaver by its heavily furred tail.

Kentucky Distribution: Statewide; common along streams.

Life History: From the mountain streams to the great reservoirs and the banks of the Mississippi River, mink can be found just about anywhere there is permanent water in Kentucky.

The best way to determine the presence of mink is to seek their characteristic footprints in the mud at the water's edge and to note droppings on a floating log, boat dock, rock, or other raised streamside object. One who is familiar with mink is sometimes alerted to their presence by the faint odor from the scent glands, which lingers wherever mink are active.

Although primarily nocturnal, mink are active morning and evening and, infrequently, in bright daylight as well. They are not particularly shy; the careful observer can expect to see several in his lifetime. They show the same curiosity as other mustelids, responding to squeaks by coming to investigate.

Mink are active at all seasons. They can easily be tracked in the morning after a snowfall. At times of extreme cold they seem somewhat less active than at other times.

The usual gait is a series of bounds that leave tracks a little more than a foot apart. Mink readily take to the water and are excellent swimmers both at and below the surface.

Distribution of *Mustela vison* in the contiguous United States

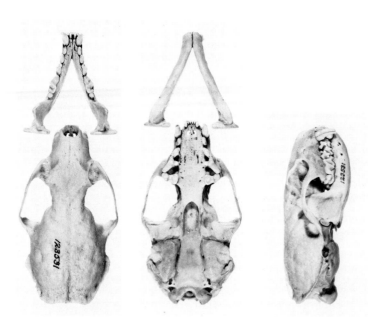

Skull of *Mustela vison*, x 0.8

The breeding season begins in January and extends into March. Implantation of the fertilized ova is delayed for varying periods. The litter of 3 to 6 is produced in April, 30 to 32 days after implantation occurs. The kits are naked, blind, and pink at birth. Their eyes open at 25 days of age, the young begin to take solid food brought in by the mother at 5 or 6 weeks, and the youngsters begin to capture some of their own food when they are 8 weeks old. The mother and young stay together until autumn, at which time the young disperse.

The den is near water, often in an old muskrat burrow. A natural rock crevice or a cavity among rocks at the base of a bridge may be used. Sometimes a mink digs its own burrow, under the roots of a streamside tree or in a similarly sheltered spot. The nest in which the young are born is made in a cavity about 3 dm in diameter and consists of grass, feathers, and fur. It is generally in a burrow some 3 m long and 6 to 10 dm below the surface. Male mink live in their own, less extensive burrows.

Food of the mink varies strikingly with the season. In summer crayfish make up most of the diet; this fare is supplemented with mammals and frogs. In winter mammals are the main course—chiefly muskrats, but rabbits and mice also. In Carter Caves State Park 2 mink found in a cave had been feeding on the bats that hibernate there.

Although birds ordinarily form a rather small part of the diet, a mink occasionally will take to raiding the henhouse and must be destroyed.

In the 5 trapping seasons 1968–69 through 1972–73 fur-buyers in Kentucky purchased 18,632 mink pelts; the number averaged 3,726 per year but varied from a low of 2,490 in 1970–71 to highs of 4,873 in 1968–69 and 4,845 in 1972–73. Average price per pelt was $6.49; the range was $4.26 in 1970–71 to $10.41 in 1972–73. The trend toward pen rearing on mink ranches in recent years has cut into the importance of trapping in the wild. Campaigns by humane societies against the cruelty of the leghold steel trap have also been a factor in the declining popularity of trapping.

Eastern Spotted Skunk;
Civit Cat P<small>LATE</small> 29

Spilogale putorius (Linnaeus)

Recognition: Total length 445–550 mm; tail 165–220 mm; hind foot 43–52 mm; weight 315–1,260 g (males larger than females). A small, slender skunk scarcely larger than a squirrel. Color black, interrupted by narrow white stripes, spots, and blotches; stripes or spots on the flanks are aligned vertically.

Variation: No geographic variation is recognized in Kentucky. Our subspecies is *S. p. putorius* (Linnaeus).

Confusing Species: Somewhat similar to the striped skunk, *Mephitis mephitis.* The striped skunk is larger and more robust and never has a spotted appearance or vertical stripes on the flanks.

Kentucky Distribution: Eastern Kentucky. This rare skunk is known from Bell, Elliott, and McCreary counties and probably occurs sparingly throughout the Cumberland Plateau.

Spotted skunks should be sought in western Kentucky. Old sight records have been perpetuated the idea that this species occurs in southern Indiana and Illinois, and the presence of a skeletal fragment in an Indian midden in the latter state suggests it has occurred there in historical times. However, there are no recent specimens from the tristate region.

Life History: The cliffs and rocks of the rugged terrain of eastern Kentucky provide shelter for the few spotted skunks in our state.

The spotted skunk is more agile than its larger relative. It moves at a bounding gait, so its footprints are farther apart (and smaller) than those of the plodding striped skunk. It climbs fences and trees. Spotted skunks have sometimes been seen climbing about on the rafters of a building.

Distribution of *Spilogale putorius* in the United States

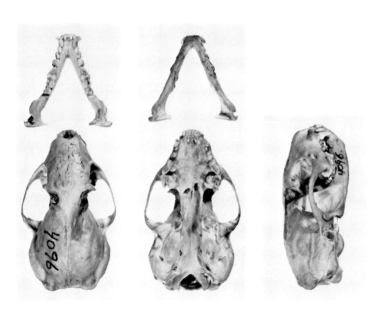

Skull of *Spilogale putorius*, x 0.9

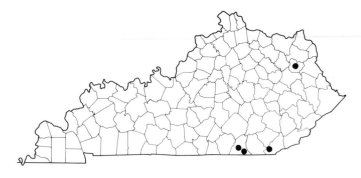

Locality records of *Spilogale putorius* in Kentucky

Spotted skunks are less active in midwinter than at other seasons, but they do not have long sleeping periods. They are nocturnal.

Like its larger cousin, the spotted skunk has an excellent defense against its enemies. It shows as much reluctance to use its spray, preferring to threaten an intruder by charging at him and then standing on its front feet. Its odor is distinctive: sharper and more pungent than that of the striped skunk.

Little is known of the breeding habits of spotted skunks. Apparently they breed in late winter and produce a litter of 2 to 6 after a gestation period of about 60 to 70 days. The young weigh only about 10 g each and are hairless, although the color pattern shows clearly on the skin. Their eyes open in 30 days. By August the youngsters are more than half-grown, but they still forage with the mother.

The den site is usually a crevice at the base of a cliff or among boulders. Grass and dry leaves are brought in to make the nest in which the young are raised.

The diet seems to be quite similar to that of the striped skunk. In summer and fall the major food is insects, especially beetles. In winter and spring mice and rabbits form the major part of the diet. Fruit, corn, birds and their eggs, and a wide variety of other items round out the menu.

The spotted skunk does well in captivity. Although it is not as docile a pet as the striped skunk, it is more active, playful, and interesting.

In the 5 trapping seasons 1968–69 through 1972–73 fur-buyers in Kentucky purchased only 8 spotted skunk pelts: 4 in 1968–69 and 4 in 1969–70. The average price of the 1968–69 pelts was $1.25; of the 1969–70 pelts, 88¢.

Striped Skunk   PLATE 30
Mephitis mephitis (Schreber)

Recognition: Total length 540–760 mm; tail 200–280 mm; hind foot 69–82 mm; weight 1.4–5.4 kg (males larger than females). A bushy-tailed, short-legged, black-and-white animal about the size of a house cat. The head is small and pointed; it is black, with a white stripe on the face.

Variation: No geographic variation is evident in Kentucky. Our subspecies is *M. m. nigra* (Peale and Palisot de Beauvois). Although the common pattern is that shown in our color plate, the amount of white on striped skunks varies tremendously. The width of the stripes commonly varies; some individuals lack stripes, being entirely black except for a patch of white on the head, and some have stripes so broad that they become fused and cover the entire back.

Confusing Species: Our only similar mammal is the spotted skunk, *Spilogale putorius*, which can be recognized by its smaller size, broken pattern of scattered spots and stripes, and vertical stripes on the flanks.

Kentucky Distribution: Statewide; common.

Life History: Farmland, brushland, woodland edge, open woods, weedy fields, rocky areas, cliffs, and small caves are all favored habitats of the striped skunk.

Distribution of *Mephitis mephitis* in the United States

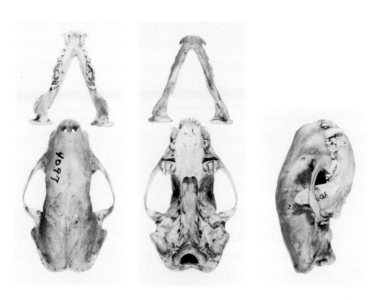

Skull of *Mephitis mephitis*, x 0.6

Home is an underground den. Often an old woodchuck burrow or abandoned fox den is taken over. If necessary, however, the skunk can dig its own; its long-clawed front feet are well adapted to digging.

Skunk dens are often associated with man. A hole beneath a barn, an abandoned house, an outbuilding or even beneath the porch of an occupied house may lead to a skunk den.

Although skunks are occasionally abroad by day, they are essentially nocturnal, emerging at dusk to forage. Often they leave signs of their nightly activity in the form of little shallow diggings where they were hunting grubs among the grass roots.

Armed with powerful scent glands, the skunk shows little fear of potential enemies; one can easily walk up to a skunk and catch it (see "Remarks") as it ambles along at a rate of about 1.5 km per hour. If pursued, a skunk can run off at a rate of about 8 km per hour.

The striped skunk is a peaceful creature and will challenge man only if cornered or threatened. As the first act of self-defense, a skunk faces the intruder, stamps its front feet, and raises its tail. It seems most reluctant to release the scent unless actually injured. In doing so, a skunk faces away from the target, raises its tail, and emits well-directed twin streams of an amber-colored fluid in a spray so fine as to be scarcely noticeable except for the overpowering odor.

Nearly half of the skunk's food consists of insects, especially grasshoppers, crickets, beetles, grubs, cutworms, bugs, bees, and wasps. Mice and other small mammals make up about 10 to 20% of the diet. Turtle eggs, dug in season from the sandy nests, are a special delicacy. Most of the rest of the food is vegetable matter. Fruit, such as dogwood, blackberry, wild grape, wild cherry and wild plum, is eaten, and a small amount of grain, grass, leaves, and buds is used in winter.

As cold weather approaches, skunks get fat and sluggish. Although they do not hibernate, they become inactive on the coldest winter days, sometimes sleeping for a week or more in a snug underground nest.

Breeding occurs in late February. On warm nights at this

Striped skunk, *Mephitis
mephitis.* Occasional individuals
have even less white.

time a skunk may wander a kilometer or more in seeking a
mate. On these forays they often encounter farm dogs or
become traffic fatalities; perhaps this is why skunk odor seems
most common on warm, rainy nights in late February and
early March.

Following a gestation period of 62 to 68 days a litter of 2
to 10 black-and-white cubs is born. They are nursed in the
den for most of their first 2 months of life, after which they
appear above ground to forage with their mother. By late
June the delightful sight of a mother with her parade of
skunklets marching in single file behind her may be seen by
the lucky observer anywhere in Kentucky.

The skunk family stays together for most of its first year.
Although some youngsters go their separate ways in the fall,
most of them den with the mother during their first winter.
Thus, a half-dozen or more skunks are sometimes found
together in a winter den.

Remarks: The persistence of skunk odor is legendary, and
many folk remedies have been devised. Turpentine is reported

to be remarkably effective. To remove skunk odor from a dog, bathe it with tincture of green soap. Wipe dry and then sponge the animal with a solution of 3 tablespoons of sodium perborate, trisodium phosphate, or a detergent in a basin of warm water. If some odor remains, sponge the animal with chlorine disinfectant, using the solution as recommended on the container. Others have reported using tomato juice or benzene to remove skunk odor. Whether any of these methods are effective we do not know.

We can confirm that one can pick up a skunk by the tail without being sprayed. It is advisable, however, to be quick and accurate in grasping the tail and to lift the animal clear of the ground almost instantaneously. To release the skunk, simply toss it gently, feet down, onto the ground a few feet in front of you, and it will scamper off without spraying.

Skunks are popular pets. They are easily de-scented by the simple operation of removing the scent glands, which are grape-sized muscular bulbs on either side just within the anal opening. Although this operation is commonly done, we think it is cruel injustice to a skunk, for it renders him helpless when attacked by dogs. A young skunk makes just as good a pet if not de-scented; it will not harm you unless mistreated.

Skunks are generally desirable animals. They destroy mice, rats, and various noxious insects. However, occasional individuals take to raiding the chickenhouse and destroy chicks and eggs. A large population of skunks can have a detrimental impact on quail and other ground-nesting birds. Along with raccoons and foxes, skunks are our most common carriers of rabies, and it is sometimes necessary to reduce the skunk population during an outbreak of this disease.

Skunk meat is reported to be delicious, with a taste and texture between pork and chicken.

Skunks have few enemies. Dogs attack them; and each dog must learn for itself about the skunk's defense. Bobcats occasionally kill and eat skunks. Great horned owls seem to be the major predator on skunks; it seems that most museum skins of this great bird smell faintly of skunk.

Parasites and disease probably are important in the natural control of skunk populations. However, man and the automobile are major factors in destroying skunks. The fearless and slow-moving skunk is one of our most frequent highway casualties.

In the 5 trapping seasons 1968–69 through 1972–73 fur-buyers in Kentucky purchased 1,254 striped skunk pelts; the average was about 251 per year, and the number varied from 173 in 1970–71 to 323 in 1968–69. Average price per pelt was 82¢; the range was 60¢ in 1970–71 to $1.07 in 1972–73.

River Otter PLATE 30

Lontra canadensis (Schreber)

Recognition: Total length 889–1,300 mm; tail 300–507 mm; hind foot 100–146 mm; weight 5–10.5 kg. A large, semiaquatic mammal with rich-brown fur having a silvery sheen below. The ears are small, the snout is broad, and the feet are webbed. The tail is long, thick at the base, and tapered toward the tip.

Variation: We have no data on present geographic variation in this species in Kentucky. Presumably the subspecies *L. c.*

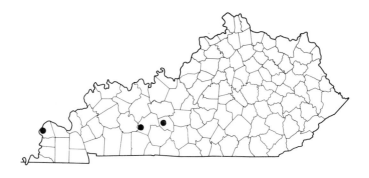

Recent locality records of *Lontra canadensis* in Kentucky

River otter, *Lontra canadensis*, in a hollow log

canadensis (Schreber) was the common river otter across most of Kentucky, being replaced in the Purchase by *L. c. interior* (Swenk), a slightly paler and somewhat larger race.

The name *Lontra* has recently been restored as the name of the genus which had been spelled *Lutra*.

Confusing Species: It is difficult to confuse this magnificent animal with any other, but it can be done. We were once sent a woodchuck carcass for positive identification, for it was thought to be an otter! The beaver, the mink, and the muskrat might be confused with the otter, but the beaver has a flat, scaly tail; the mink is smaller, has a white spot on the chin, and does not have fully webbed feet; and the muskrat has a slender, scaly tail that is flattened from side to side.

Kentucky Distribution: At one time the otter was widely distributed in Kentucky; its former distribution and abundance is attested to by the numerous "Otter Creeks" in the state. The pressures of civilization and the demand for its durable

Distribution of *Lontra canadensis* in the contiguous United States

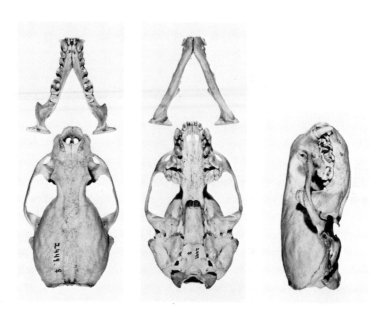

Skull of *Lontra canadensis*, x 0.4

pelt have made the river otter now extremely rare and local. There are still a few scattered individuals in the swamps, bayous, and sluggish streams of the Purchase. Perhaps they still occur in other areas, especially along the lower reaches of the streams flowing into the Ohio River below Owensboro.

Life History: The river otter is a creature of the wildlands in Kentucky; this, coupled with its rarity and shyness, renders it nearly exempt from the attention of man. We know of but 3 sightings in the last 15 years, and one of these was almost surely of an escaped animal.

An excellent swimmer, the otter progresses in an undulating course, often with all but the head submerged. It is a superb diver, able to swim for over 400 m without coming up for air. Its agility is such that it can outswim and capture a trout in open water.

Mating apparently occurs in late winter or early spring; the gestation period is 8 or 9 weeks. The 1 to 3, but usually 2, young are born in late April or early May. The natal chamber is normally in a burrow in a streambank, beneath the spreading roots of a sheltering tree.

Like other mustelids, the young are born in a rather immature state, with their eyes closed. It is nearly 2 months before they take to the water, where the mother teaches them to swim. For their lessons they climb on her back, and she swims out to deep water with them and submerges, leaving them adrift. If they get into trouble she rescues them—and repeats the process until, quite soon, they are swimming on their own. There is some evidence that the youngsters remain with the mother until winter, when they go their separate ways.

Usually about half the diet of the river otter is fish; the rest consists of frogs, crayfish, insects, and various other animals.

The only sign of otters that most of us are ever likely to see in the wild is their famous slides. When snow and ice covers the ground, an otter will climb to the top of a steep bank sloping into a deep pool, tuck his forelegs beneath him, and toboggan down the slope on his belly into the water. Apparently the tobogganing is great sport, for an otter will repeat it again and again, sometimes in company. In places or times of little or no snow, a slippery mudbank serves the same purpose.

Family Felidae   Cats

Digitigrade; toes 5 on the forefoot, 4 on the hind foot; claws compressed and retractile. Canine teeth long and sharp; large molars with shearing crowns. Represented in Kentucky by a single species, *Lynx rufus*, the bobcat. In earlier times *Felis concolor*, the panther, occurred in the state. It was readily separable from the bobcat on the basis of size (up to 2.13 m for the panther, 0.9 m for the bobcat) and tail length (178 mm for the bobcat, up to 812 mm for the panther).

Bobcat; Wildcat   PLATE 31

Lynx rufus (Schreber)

Recognition: Total length 800–1,015 mm; tail 130–180 mm; hind foot 155–197 mm; weight 6.7–16 kg. Similar to an overgrown house cat with a short tail. Upperparts gray to brown, spotted and blotched with black. Underparts whitish with blotches of black and with a conspicuous black bar on the inside of the front legs.

Variation: No geographic variation is evident in Kentucky. Our subspecies is *L. r. rufus* (Schreber).

Confusing Species: Only the house cat is similar. The bobcat can be distinguished by its short tail, measuring about a fourth of the body length.

Bobcat droppings on a log

Kentucky Distribution: Statewide; scarce.

Life History: The bobcat roams the wilder areas throughout the state, from the woodlands, brushy hollows, and cliff country of the Cumberland Mountains to the cypress swamps along the Mississippi.

This shy nocturnal hunter is seldom seen by man, but occasionally one is jumped from its daytime shelter in the grass or beneath a shrub. Rarely, a bobcat is seen hunting by day.

The best way to find evidence of bobcats is to seek droppings along a trail or its tracks in the snow. Once alerted to wildcat droppings, the outdoorsman can expect to find them with some regularity, even in wild areas close to human habitation.

Bobcats use trails and roads in their nightly travels. Droppings are usually deposited on a rise in the trail, on a rock or log, or on some other elevated place. They consist almost entirely of hair and bones—even more so than those of foxes; and they tend to survive much weathering. An experienced observer can sometimes find signs of the cats by walking a few hundred yards along a woodland path in suitable areas.

As active in winter as summer, bobcats can be located by their tracks in the snow. The footprint covers twice the area of a house cat's print but otherwise is similar.

Breeding occurs in February or March. After a gestation period of about 62 days a litter of 1 to 4 (usually 2) kittens is born in a crude nest of leaves and moss in a rock crevice, hollow tree, or similar shelter. The eyes open when the kittens

Distribution of *Lynx rufus* in the United States

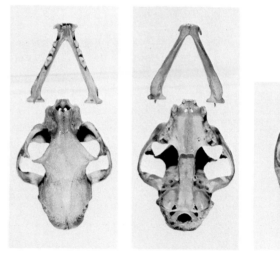

Skull of *Lynx rufus*, x 0.3

are 9 or 10 days old. They are weaned at 60 to 70 days of age but tend to stay with their mother into autumn.

Rabbits are the major food item, often making up half or more of the diet. Mice, muskrats, opossum, and grouse are eaten. Deer form a major part of the diet during and just after the hunting season, when cripples and lost kills are available. Occasionally a wildcat will take a fawn, and there is now an authentic record of one pulling down a healthy adult deer.

Bobcats tend to stay away from active farms, but occasionally one will kill a young calf, a lamb, or a chicken.

Bobcats hunt like the house cat: by stealth or by lying in wait to pounce upon the prey. An attempt to catch a rabbit calls for a short burst of speed; if the rabbit is not caught at once, it is not pursued.

Where bobcats are scarce, as in Kentucky, they are neither beneficial nor harmful to man. They are an interesting part of our natural fauna, and we can be thankful to have them.

In the 5 trapping seasons 1968–69 through 1972–73 fur-buyers in Kentucky purchased 24 bobcat skins; although the average was about 5 per year, purchases varied from none in 1970–71 to 10 in 1972–73. Average price per pelt was $3.96; the range was $3.33 in 1968–69 to $8.65 in 1972–73.

ORDER ARTIODACTYLA *Even-toed Ungulates*

Members of this order have the main axis of the foot passing between the third and fourth toes. In most, there is complete suppression of all other toes, leaving but these 2; in the others, the inner (first) digit only is suppressed, leaving 4 functional toes. The molars are broad-crowned; they differ from the premolars in having 2 crowns. Artiodactyls usually are of large size, and most of them are strictly herbivorous. The order is essentially worldwide in distribution (absent from the Australian region). There are about 170 species, arranged in 9 families. The 2 species in Kentucky are members of the family Cervidae. In this family the males (and, rarely, the females) have bony antlers. There are no upper incisors, and the upper canines are usually absent. There are 3 lower incisors on each side, with an incisorlike canine in contact with them.

KEY TO SPECIES OF KENTUCKY ARTIODACTYLA

1. a. Tail black above; summer pelage usually with large white spots on the back and sides; uniform gray in winter; antlers palmate (flattened, with tines): *Cervus dama*, Fallow Deer, p. 282
   b. Tail not black above; summer pelage reddish; winter pelage bluish gray; antlers not palmate: *Odocoileus virginianus*, White-tailed Deer, p. 286

Fallow Deer   PLATE 32
Cervus dama Linnaeus

Recognition: Total length 1,450–1,700 mm; tail 160–190 mm; weight 31–100 kg. A medium-sized deer, about 900 to 1,000 mm tall at the shoulder. The pelage is more or less white-spotted. Tail black above, white below. Antlers palmate, with brow tines.

Variation: No geographic variation is recognized in Kentucky, but there may be great variation from one individual to another. There is also seasonal variation: in summer the pelage is white-spotted and often has large, irregular white blotches on the back and sides; in winter the spots become less conspicuous, and the color becomes a nearly uniform gray.

Confusing Species: Similar to the white-tailed deer, *Odocoileus virginianus*, but the antlers are palmate (like a hand) and

Fallow deer, *Cervus dama*

have brow tines. Those of the white-tailed deer are sticklike, and brow tines are lacking. Does and antlerless bucks of the 2 species are more difficult to distinguish. Fallow deer are chestnut with white spotting in summer and are gray in winter. White-tailed deer are tan in summer and, except for the fawns, are without spots; they are bluish-gray in winter. The top of the tail of the fallow deer is black; that of the white-tailed deer is about the same color as the back but is never black. In both species the underside of the tail is white.

Kentucky Distribution: Common in the Land Between the Lakes, in Lyon and Trigg counties. Introduced, with varying success, usually poor, in some other areas; some still survive in Madison County and around Camp Webb, at Grayson Reservoir, in eastern Kentucky.

Life History: This species is a native of Asia Minor and the Mediterranean countries. Because they thrive in captivity and yield excellent venison, they have been a favorite park deer for centuries; they were apparently introduced into Great Britain by the Romans.

Fallow deer are thoroughly gregarious. However, summer herds consist almost wholly of females and immature males, the adult males remaining apart.

In late April or early May most bucks shed their antlers— the older animals earlier than the young; sometimes young bucks retain their antlers into June. Antler growth starts immediately after shedding, and by early August the new antlers are nearly fully formed. By late August the bucks are rubbing the velvet from the antlers, and thereafter the bucks are much in evidence as they gather into small herds composed solely of animals of their own sex.

By mid-September the bucks begin to wander, seeking territories for the rut, which occurs in mid-October. Each buck selects one or more rutting territories, which he marks by rubbing the secretions of his scent glands (located below the eyes) on trees and shrubs and by urinating on bare places he scrapes on the ground.

By this time the neck has greatly thickened and the Adam's apple is more pronounced. By early October the buck begins to utter a call heard only in the rutting season: a throaty, grunting groan, sometimes almost a belching sound, given over and over. Young bucks have higher-pitched voices than older ones. An old buck can sometimes be heard at a distance of more than 0.8 km. If disturbed while calling, he will retire from the area, but he will return shortly or else go to a previously selected and marked alternative area.

The rutting territory is roughly 0.2 hectare in extent, and the buck parades back and forth there, calling almost continuously for hours on end. The does are attracted to the territory by both scent and sound and are served as they come in season, sometimes after a lengthy chase.

By the end of October the rut is essentially over, and the exhausted buck deserts his territories to hide away in the woods. Bucks are little in evidence at this time, but in a few weeks

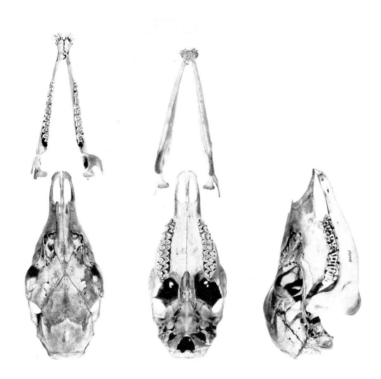

Skull of *Cervus dama,* x 0.2

they begin to congregate in small, all-male herds. Very old bucks may remain solitary.

Does do not normally breed until they are 28 months old. The gestation period is 8 months. The single fawn (twins are very rare) is usually born in mid-June, and it normally runs with the mother until the following March.

Fallow deer eat a great variety of grasses, herbs, bushes, fruits, and nuts. Acorns are a favorite food, and in fall many varieties of fungi are eaten.

Remarks: The fallow deer is considered a game species in Kentucky, and is hunted by permit in the Land Between the Lakes. In the past 5 years (1969–1973) there has been a gradual increase in the take, and the average annual harvest is 95 animals.

White-tailed Deer   Plate 32

Odocoileus virginianus (Zimmermann)

Recognition: Total length 1,600–2,150 mm; tail 255–360 mm; hind foot 480–538 mm; weight 40–135 kg (males larger than females). A large hoofed animal, with a rather long tail that appears as a conspicuous white flag as the animal flees. Color is tan above in summer, blue-gray in winter; nearly white below. Fawns are white-spotted. Adult males have conspicuous antlers most of the year; the antlers are shed in midwinter, but soon grow again.

Variation: No geographic variation is evident in Kentucky. Our native subspecies is *O. v. virginianus* (Zimmermann).

Confusing Species: Similar only to the introduced fallow deer, which is recognized by its smaller size, palmate antlers, and spotted coat. For the differences between does and antlerless bucks of the 2 species see the preceding account.

Kentucky Distribution: Statewide. The species varies from fairly common in many of the less settled counties to scarce and local in heavily settled regions and in the Bluegrass.

Life History: The deer is an animal of second-growth woodland and of forest edge, brushland, and open forest. Its favored habitat is timber slash, where selective cutting of large timber trees has allowed the growth of numerous small trees and herbs, which provide good food and cover.

The presence of deer is easily detected by the distinctive hoof-prints and the piles of small droppings, which are almost spherical and about 12 mm in diameter. If you are fortunate, when walking through the brushland you may rouse a small group of deer, hear them go crashing off through the brush, and catch a glimpse of the white flags of their tails as they flee.

Although deer are primarily nocturnal, they are rather frequently seen by day in places where they are abundant, such as Mammoth Cave National Park. Early morning and evening are the best times to look for them.

A feeding deer wanders slowly and quietly, browsing on the twigs of trees and shrubby vegetation. They do little grazing but do take some clover and other broad-leaved herbaceous plants. A deer consumes 2.7 to 3.6 kg of browse per day.

Deer are remarkably sedentary for such large animals, spending most of their lives in an area no more than 1.6 km across.

When frightened a deer bounds off at a speed of 48 km per hour or greater. It is an excellent jumper, easily clearing a 2 m fence. If pressed it will take to the water; deer swim well.

The mating, or rutting, season is November and December. At this time bucks become aggressive and fight violently among themselves, using both hoofs and antlers. Sometimes a combatant is killed, and occasionally two bucks will lock antlers so firmly that they cannot separate, and both die.

Gestation takes about 196 days, and 1 or 2 fawns are born, in May. The fawns, weighing 2.5–4 kg at birth, are delightfully gentle, spotted creatures. In their early days they are left alone,

White-tailed deer, *Odocoileus virginianus*, fawn

bedded down in a secluded spot, while the mother goes off to feed. If you find one, let it be; the mother, though you will not see her, has not deserted it.

The white spots disappear with the fawn's first molt, which occurs in its fourth or fifth month. The young doe does not breed in her first winter; she produces her first fawn when she is 2 years old.

The critical time in the life of white-tailed deer is the winter season, when they put a heavy demand on a browse supply that is not growing. In a heavy snow, travel becomes difficult, and deer tend to congregate and feed in a single area, which is referred to as a "deer yard." As food becomes scarce, the deer tend to starve rather than move out. Fortunately, snows in Kentucky are rarely so heavy as to cause the deer to yard.

Although deer are highly desirable animals, prized for aesthetic reasons and as game, large populations present problems and give rise to highly emotional controversies. Formerly the panthers, wolves, and other large predators controlled the population, but their elimination in most of the eastern states left up to man the responsibility of managing the numbers of deer. Where there are large tracts of favorable habitat, it is necessary to have well-planned hunting seasons (sometimes including a kill of does) in order to prevent overpopulation, destruction of the food supply, and massive winter starvation. Although these classic deer problems occur primarily in the northern states, we have them sometimes in Mammoth Cave National Park and other localities.

The white-tailed deer is the prize game animal of Kentucky. Although our herd is small in comparison with those of the northern states or even West Virginia, it provides sport for more than 45,000 hunters. The harvest is rather stable at about 8,500 per year. The number of hunters, however, has grown so much that the kill per hunter has been halved since 1967.

When deer populations are high, considerable damage can be done to crops. Deer destroy farm and garden produce, small orchard trees, and nursery stock.

Deer are commonly harassed and killed by packs of free-

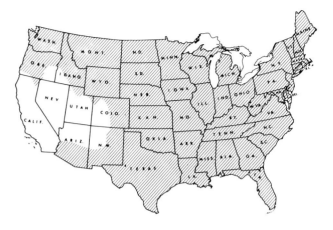

Distribution of *Odocoileus virginianus* in the United States

Skull of *Odocoileus virginiana*, x 0.2

White-tailed deer,
Odocoileus virginianus, doe

running dogs. This is a major factor limiting their population in most of Kentucky and in preventing their establishment in many otherwise suitable localities. Throughout the eastern United States deer are abundant only in areas where people and their dogs are scarce or where swampland provides a sanctuary from the dogs. Drainage of swamps and the channelization of streams, so extensively practiced by the U.S. Soil Conservation Service and the U.S. Army Corps of Engineers in western Kentucky and throughout the South, pose a serious threat to the survival of deer in these areas.

Deer commonly fall victim to the automobile—and the automobile and its passengers can, at the same instant, fall victim to the deer, in a wreck caused by striking so large an animal. The fences along interstate highways are no barrier to a deer.

Not infrequently a deer will walk onto the ice on a pond or reservoir, break through, and drown.

Dental Formulae
of Kentucky Mammals

Expressed as the number of teeth per side. Numerator = upper teeth, denominator = lower teeth.

	Incisors	Canines	Premolars	Molars	Total Teeth
ORDER MARSUPIALIA					
FAMILY DIDELPHIDAE					
Didelphis	5/4	1/1	3/3	4/4	50
ORDER INSECTIVORA					
FAMILY SORICIDAE					
Sorex, Blarina	3/1	1/1	3/1	3/3	32
Cryptotis	3/1	1/1	2/1	3/3	30
FAMILY TALPIDAE					
Parascalops	3/3	1/1	4/4	3/3	44
Scalopus	3/2	1/0	3/3	3/3	36

ORDER CHIROPTERA
FAMILY VESPERTILIONIDAE

Myotis	2/3	1/1	3/3	3/3	38
Lasionycteris	2/3	1/1	2/3	3/3	36
Pipistrellus	2/3	1/1	2/2	3/3	34
Eptesicus	2/3	1/1	1/2	3/3	32
Lasiurus	1/3	1/1	2/2	3/3	32
Nycticeius	1/3	1/1	1/2	3/3	30
Plecotus	2/3	1/1	2/3	3/3	36

FAMILY MOLOSSIDAE

Tadarida	1/2 or 1/3	1/1	2/2	3/3	30 or 32

ORDER LAGOMORPHA
FAMILY LEPORIDAE

Sylvilagus	2/1	0/0	3/2	3/3	28

ORDER RODENTIA
FAMILY SCIURIDAE

Tamias	1/1	0/0	1/1	3/3	20
Marmota	1/1	0/0	2/1	3/3	22
Sciurus carolinensis	1/1	0/0	1/1 or 2/1	3/3	20 or 22

	Incisors	Canines	Premolars	Molars	Total Teeth
Sciurus niger	1/1	0/0	1/1	3/3	20
Glaucomys	1/1	0/0	2/1	3/3	22
FAMILY CASTORIDAE					
Castor	1/1	0/0	1/1	3/3	20
FAMILY CRICETIDAE					
Oryzomys, Reithrodontomys, Peromyscus, Ochrotomys, Sigmodon, Neotoma, Clethrionomys, Microtus, Ondatra, Synaptomys	1/1	0/0	0/0	3/3	16
FAMILY MURIDAE					
Rattus, Mus	1/1	0/0	0/0	3/3	16
FAMILY ZAPODIDAE					
Zapus	1/1	0/0	1/0	3/3	18
Napaeozapus	1/1	0/0	0/0	3/3	16
ORDER CARNIVORA					
FAMILY CANIDAE					
Canis, Vulpes, Urocyon	3/3	1/1	4/4	2/3	42

FAMILY URSIDAE					
Ursus	3/3	1/1	4/4	2/3	42
FAMILY PROCYONIDAE					
Procyon	3/3	1/1	4/4	2/2	40
FAMILY MUSTELIDAE					
Mustela, Spilogale, Mephitis	3/3	1/1	3/3	1/2	34
Lontra	3/3	1/1	4/3	1/2	36
FAMILY FELIDAE					
Lynx	3/3	1/1	2/2	1/1	28
ORDER ARTIODACTYLA					
FAMILY CERVIDAE					
Cervus, Odocoileus	0/3	0/1	3/3	3/3	32

APPENDIX II

Key to the Skulls
of Kentucky Mammals

Including Some Common Domestic Species

1. a. Incisors chisel-like, not more than 2 in lower jaw; canines absent: 2
   b. Incisors, if present, not chisel-like: 3

2. a. Upper incisors 4, the second pair small and peglike, directly behind the first pair; maxilla incompletely ossified: *LAGO-MORPHA*
   b. Only 2 incisors in upper jaw; maxilla completely ossified: *RODENTIA*

3. a. Teeth 50; a high median ridge on the skull; mandible with a unique inward-turning process at the rear: *MARSUPI-ALIA*
   b. Teeth fewer than 50; no high median ridge on the skull; mandible without an inward-turning process: 4

4. a. Upper incisors and canines absent: *ARTIODACTYLA* (in part)
   b. Upper incisors and canines present: 5

5. a. Skull more than 225 mm in length: 6
   b. Skull less than 225 mm in length: 8

6. a. Molars tuberculate, rounded, or oval in cross-section; canines prominent: 7
   b. Molars with ridges of enamel, alternating with bands of dentine, and squarish or rectangular in cross-section; canines sometimes absent: *PERISSODACTYLA*

7. a. Upper canines triangular in cross-ection, flaring laterally;
      skull elongate: *ARTIODACTYLA* (in part)
   b. Upper canines round in cross-section and pointing down-
      ward; skull massive: *CARNIVORA* (in part)

8. a. Canines almost the same size as the other teeth; zygomatic
      arch weak or absent: *INSECTIVORA*
   b. Canines distinctly larger than the other teeth; zygomatic
      arch well developed: 9

9. a. Skull less than 25 mm in length; incisors 1/3 or 2/3; anterior
      end of the skull broad and blunt, with a decided U-shaped
      notch: *CHIROPTERA*
   b. Skull more than 25 mm in length; incisors 3/3; anterior
      end of the skull not as above: *CARNIVORA* (in part)

KEYS TO GENERA AND SPECIES

ORDER MARSUPIALIA

Represented in the state by a single species, *Didelphis virginiana*.

ORDER INSECTIVORA

1. a. Length more than 30 mm; teeth unpigmented; zygomatic
      arch complete: 2
   b. Length less than 30 mm; tips of the teeth pigmented,
      brownish; zygomatic arch incomplete: 3

2. a. Eight teeth on each side of the lower jaw; length more than
      34 mm, width more than 17 mm: *Scalopus aquaticus*
   b. Eleven teeth on each side of the lower jaw; length less than
      34 mm; width less than 16 mm: *Parascalops breweri*

3. a. Three unicuspids visible from the side: *Cryptotis parva*
   b. Four or 5 unicuspids visible from the side: 4

4. a. Length 20 mm or more; palate length more than 9 mm; 4
      unicuspids visible from the side; fifth unicuspid minute and
      inside the tooth row; 7–8 mm across last molars: *Blarina
      brevicauda*

b. Length less than 20 mm; palate length less than 9 mm; 5
     unicuspids visible from the side; 4–5 mm across last molars:
     5

5. a. Length 17.5–18 mm; width 8.5–9 mm: *Sorex fumeus*
   b. Length less than 17 mm; width less than 8.5 mm: 6

6. a. Length 15.1–16.5 mm; width 7.3–8 mm; maxillary tooth
      row 5.5 mm; third unicuspid larger than the fourth: *Sorex
      cinereus*
   b. Length to 14.5 mm; width to 7.3 mm; maxillary tooth row
      5 mm; third unicuspid considerably smaller than the fourth:
      Sorex longirostris

ORDER CHIROPTERA

1. a. One upper incisor on each side: 2
   b. Two upper incisors on each side: 6

2. a. One upper premolar on each side; teeth 30: *Nycticieus
      humeralis*
   b. Two upper premolars on each side; teeth 30 or 32: 3

3. a. Two lower incisors on each side; teeth 30: *Tadarida brasil-
      iensis*
   b. Three lower incisors on each side; teeth 32: 4

4. a. Space between the upper incisors less than ¼ the length of
      the upper tooth row: *Tadarida brasiliensis*
   b. Space between the upper incisors more than ¼ the length
      of the upper tooth row (*Lasiurus*): 5

5. a. Length less than 15 mm; lower tooth row 5–5.5 mm long;
      upper tooth row 5 mm long: *Lasiurus borealis*
   b. Length more than 15 mm; lower tooth row 6–6.5 mm long;
      upper tooth row 6 mm long: *Lasiurus cinereus*

6. a. One upper premolar on each side; upper and lower tooth
      rows each 8 mm long or more; length 19–20 mm: *Eptesicus
      fuscus*
   b. Two or 3 upper premolars on each side; upper and lower
      tooth rows each 7 mm long or less; length less than 17 mm:
      7

7. a. One tiny tooth just behind each upper canine: 8
   b. Two tiny teeth just behind each upper canine: *Myotis* species*

8. a. Length more than 15 mm; upper and lower tooth rows each more than 6 mm long: 9
   b. Length less than 15 mm; lower tooth row 5.5–6 mm long; upper tooth row 5–5.1 mm long: *Pipistrellus subflavus*

9. a. Upper outline of the skull nearly flat: *Lasionycteris noctivagans*
   b. Upper outline of the skull decidedly curved: 10

10. a. First upper incisor unicuspid: *Plecotus townsendii*
    b. First upper incisor bifid: *Plecotus rafinesquii*

ORDER LAGOMORPHA

1. a. Length more than 80 mm, width more than 37.5 mm: *Sylvilagus aquaticus*
   b. Length less than 80 mm, width less than 37.5 mm: 2

2. a. Supraorbitals small; posterior processes short, tapering posteriorly to a slender point, free from or barely touching the skull and narrowing anteriorly until the anterior process and notch are usually absent: *Sylvilagus transitionalis*
   b. Supraorbitals broadly developed; posterior process usually broadly strap-shaped and coalescing with the skull posteriorly or, sometimes, along its entire length; anterior process broad and commonly extended to the nearly closed anterior notch: *Sylvilagus floridanus*

ORDER RODENTIA

1. a. Cheek teeth 3/3 or 4/3: 2
   b. Cheek teeth 5/4 or 4/4: 19

2. a. Molars flat-crowned, with crescents or with loops and triangles on the grinding surface; incisors usually broad: 3

* For distinguishing characteristics in this difficult genus see species accounts (pp. 60-85). Doubtful specimens should be submitted to an expert for identification.

b. Molars tuberculate, without a pattern of loops and triangles; incisors narrow: 10

3. a. Length more than 30 mm: 4
   b. Length less than 30 mm: 6

4. a. Interorbital region with a median crest; molars with triangles on the grinding surface; zygomata broadly strap-shaped: *Ondatra zibethicus*
   b. Interorbital region without a median crest; molars with crescentic loops; zygomata narrow: 5

5. a. Length more than 40 mm: *Neotoma floridana*
   b. Length less than 40 mm: *Sigmodon hispidus*

6. a. Upper incisors with a shallow groove on the anterior face: *Synaptomys cooperi*
   b. Upper incisors not grooved: 7

7. a. Posterior border of the palate ending in a shelflike structure without a median projection, connecting with the palatines only at the sides of the narial cavity; molars rooted; anterior faces of molars usually flat or with a central depression: *Clethrionomys gapperi*
   b. Posterior border of the palate with a median projection, connecting with the palatines at the center as well as at the sides of the narial cavity; molars rootless; anterior faces of molars usually rounded, with no central depression: 8

8. a. Anterior palatine foramina more than 4.5 mm in length; interorbital width less than 4 mm: 9
   b. Anterior palatine foramina less than 4 mm in length; interorbital width more than 4 mm: *Microtus pinetorum*

9. a. Crown of the second upper molar with 4 irregular loops: *Microtus ochrogaster*
   b. Crown of the second upper molar with 4 triangles and a posterior loop: *Microtus pennsylvanicus*

10. a. Upper incisors grooved on the anterior face: 11
    b. Upper incisors not grooved: 13

11. a. Length less than 20 mm, width less than 10 mm; infraorbital foramina small, about 1 mm; interorbital constriction less than 3 mm: *Reithrodontomys humulis*
    b. Length more than 20 mm, width more than 10 mm; infraorbital foramina large, about 3 mm; interorbital constriction greater than 3 mm: 12

12. a. Cheek teeth 4/3: *Zapus hudsonius*
    b. Cheek teeth 3/3: *Napaeozapus insignis*

13. a. Molars with tubercles in 3 longitudinal series: 14
    b. Molars with tubercles in 2 series: 15

14. a. Length more than 25 mm: *Rattus norvegicus*
    b. Length less than 25 mm: *Mus musculus*

15. a. Length more than 28 mm: 16
    b. Length less than 28 mm: 17

16. a. Length less than 30 mm: *Peromyscus gossypinus*
    b. Length 30 mm or more: *Oryzomys palustris*

17. a. Palatine slits essentially parallel-sided; little bulge of the maxilla in front of the infraorbital foramina; rostum narrow: 18
    b. Palatine slits bowed out in the middle; a considerable bulge of the maxilla in front of the infraorbital foramina; rostum shorter and stouter: *Peromyscus leucopus*

18. a. Length usually less than 25.5 mm, width 13.5 mm or less; palate extending to the back edge of the hindmost molar: *Peromyscus maniculatus*
    b. Length usually more than 25.5 mm, width 13.5 mm or more; palate not extending to the back edge of the hindmost molar: *Ochrotomys nuttalli*

19. a. Cheek teeth 4/4: 20
    b. Cheek teeth 5/4: 23

20. a. Length more than 80 mm: *Castor canadensis*
    b. Length less than 80 mm: 21

21. a. Length more than 65 mm: *Sciurus niger*
    b. Length less than 65 mm: 22

22. a. Length 55–65 mm: *Sciurus carolinensis*
    b. Length 40–45 mm: *Tamias striatus*

23. a. Length more than 68 mm, width more than 40 mm; skull flattened dorsally; incisors usually white on the anterior surface: *Marmota monax*
    b. Length less than 68 mm, width less than 40 mm; skull convex dorsally; incisors usually yellow or orange on the anterior face: 24

24. a. Length 34–36 mm; interorbital region less than 10 mm across and deeply notched: *Glaucomys volans*
    b. Length more than 40 mm; interorbital region more than 10 mm across and not deeply notched: *Sciurus carolinensis*

ORDER CARNIVORA

1. a. Larger upper molar teeth with crowns of the crushing type, without cutting edges on the outer side: 2
   b. Larger upper molar teeth with crowns at least partly of the cutting type, especially on the outer side: 3

2. a. Skull large, more than 250 mm long; 3 lower molars on each side; teeth 42: *Ursus americanus*
   b. Skull smaller, less than 150 mm long; 2 lower molars on each side; teeth 40: *Procyon lotor*

3. a. Molar teeth all of the cutting type (no crushing surfaces): 4
   b. Molar teeth, at least the last, with a crushing surface in addition to the cutting edge; cheek teeth 4/5, 5/5, or 6/7: 5

4. a. Cheek teeth 3/3; teeth 28: *Lynx rufus*
   b. Cheek teeth 4/3; teeth 30: *Felis domesticus* (House Cat)

5. a. Cheek teeth 6/7; rostrum long: 6
   b. Cheek teeth 4/5 or 5/5; rostrum relatively short: 9

6. a. Length more than 160 mm: 7
   b. Length less than 160 mm: 8

7. a. Rostral width less than 18% of skull length: *Canis latrans*
   b. Rostral width more than 18% of skull length: *Canis familiaris* (Domestic Dog)

8. a. Cranial ridges forming a narrow V and forming a sagittal crest posteriorly; sides of the braincase smooth: *Vulpes vulpes*
   b. Cranial ridges forming a broad U and not forming a sagittal crest posteriorly; sides of the braincase rough: *Urocyon cinereoargenteus*

9. a. Cheek teeth 5/5; skull nearly as wide at the middle of the braincase as at the zygomatic arch: *Lontra canadensis*
   b. Cheek teeth 4/5; skull shape not as above: 10

10. a. Median lobe of the palate not extending appreciably beyond the posterior edges of the last upper molars; last upper molar large and quadrangular: 11
    b. Median lobe of the palate extending appreciably beyond the posterior edges of the last upper molars; last upper molar transverse or almost triangular: 12

11. a. Length to 50 mm, width about 35 mm: *Spilogale putorius*
    b. Length 60–65 mm, width 36–40 mm: *Mephitis mephitis*

12. a. Length about 70 mm: *Mustela vison*
    b. Length 40–48 mm: *Mustela frenata*

ORDER PERISSODACTYLA

Although there are no native members of this order—the odd-toed ungulates—in Kentucky, there is an abundance of domesticated representatives: horses and their kin (ponies, donkeys, mules).

ORDER ARTIODACTYLA

1. a. Upper incisors and upper canines present: *Sus scrofa* (Pig)
   b. Upper incisors and upper canines absent: 2

2. a. Length more than 260 mm: *Bos taurus* (Cattle)
   b. Length less than 260 mm: 3

3. a. Permanent horn cores present in both sexes: 4
   b. No permanent horn cores: 5

4. a. Anterior face of the first upper premolar with a vertical groove; nasals strongly bowed downward: *Ovis* (Sheep)
   b. No vertical groove on the anterior face of the first upper premolar; nasals not as strongly bowed: *Capra* (Goat)

5. a. Rostrum short and broad, its length anterior to the first premolar less than 30% of the total length of the skull: *Cervus dama*
   b. Rostrum long and slender, its length anterior to the first premolar more than 30% of the total length of the skull: *Odocoileus virginianus*

APPENDIX III

Mammals of Problematic or Relatively Recent Occurrence in Kentucky

In a recent paper, Guilday, Hamilton, and McCrady (1971) listed a number of mammal species identified from recent fossil remains in Welsh Cave, in Woodford County. Carbon dating indicated that the remains were approximately 13,000 years old; i.e., the animals were here in late Pleistocene (Ice Age) times. Some of the species found in Welsh Cave are living in Kentucky today. Others, such as the dire wolf, *Canis dirus*, the mammoth, *Mammuthus* and a peccary, *Platygonus compressus*, are extinct. A few others have completely disappeared from this area; they exist today nowhere near Kentucky. These include the spruce vole, *Phenacomys*, the yellow-cheeked vole, *Microtus xanthognathus*, the porcupine, *Erethizon dorsatum*, the grizzly bear, *Ursus arctos horribilis*, and the badger, *Taxidea taxus*.

From the standpoint of the present mammalian fauna of Kentucky, however, the most interesting of the Welsh Cave finds are 7 species not now known to live in Kentucky but occurring within 150 km or less of our boundaries. These 7 species are included in the following annotated list, and each is marked with an asterisk (*).

There are other published or oral reports of the occurrence in Kentucky of a few species of mammals not treated in this volume. They have been excluded because the species are no longer here, because the reports are erroneous, or because the reports are not based on sufficient evidence to cause us to consider the animal a valid member of our fauna.

Other species, not yet found in Kentucky, may in fact be here. Still others are extending their range in our direction and at such a rate as to cause us to expect them to occur in Kentucky shortly. Also, man has been responsible, both by accident and by design, for the importation of various species. Exotic bats may wander into Kentucky or may be carried here by storms.

The following annotated list of 21 species includes those we consider of possible, or perhaps likely, occurrence.

Sorex dispar Batchelder. LONG-TAILED SHREW. Similar in appearance to the smoky shrew, *S. fumeus*, but slimmer, with a longer tail, and nearly uniformly colored, the belly being almost as dark as the back. This northern species occurs on the higher peaks of the Appalachians south to North Carolina and Tennessee; perhaps it also occurs in isolated spots on the higher mountains of southeastern Kentucky.

Sorex palustris Richardson. WATER SHREW. This large, semi-aquatic, dark-backed, light-bellied, long-tailed shrew may best be recognized by its broad hind feet, which are conspicuously fringed with stiff hairs. This is a northern species, one race of which extends southward along the Appalachian crest into eastern Tennessee.

Microsorex thompsoni (Baird). PIGMY SHREW. This tiny, long-tailed shrew can be readily distinguished from our other long-tailed shrews by the fact that it has only 3 unicuspids visible from the side; others have 5. There is a dried carcass of this species in the collections of the University of Kentucky. The label bears the following notation: "Ky., Dec. 30, 1904. *S. hoyi?* Miller." Presumably, the specimen was collected by Dr. Arthur M. Miller, a member of the staff of the University of Kentucky from 1892 to 1929. According to Dr. William S. Webb, who knew Dr. Miller intimately, the handwriting on the label is that of Dr. Miller. It has been established that Dr. Miller, although a geologist by profession, was an able collector and had more than a passing interest in vertebrate animals, and particularly in shrews. It is unfortunate that a more accurate record of the locality whence the specimen came is unobtainable. However, the specimen stands as evidence of the recent occurrence of the species within the state.

Condylura cristata (Linnaeus). STAR-NOSED MOLE. This mole is unique in having a fringe of 22 fleshy tentacles surrounding the tip of the snout. This northern species ranges widely just to the north and the east of Kentucky, so it may occur here. We have oral reports of the species from Jessamine and Fayette counties, but there are no specimens extant. In spite of much searching, we have not been able to find the species in Kentucky.

Lasiurus seminolus (Rhodes). SEMINOLE BAT. Very similar to the red bat, *Lasiurus borealis*, but deep mahogany in color. This species of the Deep South sometimes wanders northward in late summer; there are records from Tennessee, New York, and Pennsylvania, and in all probability it occasionally enters Kentucky.

Lasiurus intermedius H. Allen. NORTHERN YELLOW BAT. Similar to our red bat but larger and clearly yellowish in color, from yellowish-orange through yellowish-brown to nearly gray. This bat of the Spanish moss of the Deep South apparently sometimes wanders northward; there is an October record from New Jersey and a May record from Norfolk, Virginia. A transient, far north of the normal range, could occur in Kentucky.

Dasypus novemcinctus Linnaeus. NINE-BANDED ARMADILLO. This peculiar animal is best characterized by its shell-like, scaly skin with 9 transversely joined bands across its back. The tail is long and stiff and the claws are long and prominent. This animal, characteristic of the south-central United States, has been extending its range eastward and northward in recent years. It now occurs northward into Colorado, eastward across Kansas into Missouri, in northern Mississippi, and eastward to Georgia and most of Florida. We should not be surprised to have this species appear in western Kentucky within a few years.

*Lepus americanus Erxleben. SNOWSHOE HARE. This large dark brown hare turns white in winter. It is a northern species, ranging southward along the crest of the Appalachians into eastern Tennessee.

*Spermophilus tridecemlineatus (Mitchill). THIRTEEN-LINED GROUND SQUIRREL. This species is readily distinguished by the 13 light stripes on the brownish back. This is an animal of the Great Plains and the upper Midwest; nearest to us, it occurs across the northern two-thirds of Illinois and Indiana and in Ohio southward almost to Cincinnati.

*Tamiasciurus hudsonicus (Erxleben). RED SQUIRREL. This bushy-tailed little red tree squirrel is about halfway between a chipmunk and a gray squirrel in size. This squirrel is characteristic of the coniferous woodlands of the north, but it ranges southward into Indiana, central Ohio, and in the Appalachians into Tennessee, North Carolina, and South Carolina.

Glaucomys sabrinus (Shaw). NORTHERN FLYING SQUIRREL. Similar to the southern flying squirrel but larger and browner and with the bases of the belly hairs dark instead of uniform white. This chiefly northern species is found along the crest of the Appalachians into North Carolina and Tennessee. It might occur in the highest parts of the southeastern mountains of Kentucky but has not been recorded.

*Geomys bursarius (Shaw). PLAINS POCKET GOPHER. The permanently exposed incisors, the fur-lined, external cheek pouches, the long, curved claws, and the naked, short tail distinguish this burrowing animal from any Kentucky species. This is a plains animal, ranging eastward in Missouri to the Mississippi River and across the river into south-central Illinois and northern Indiana.

Reithrodontomys megalotis (Baird). WESTERN HARVEST MOUSE. This species is quite similar to the eastern harvest mouse but differs in being browner and having a lighter belly. This is a western species that is extending its range eastward. It has been recorded in northern Indiana and in eastern Arkansas, only 100 km from the Kentucky border, so it may someday arrive in western Kentucky.

Reithrodontomys fulvescens J. A. Allen. FULVOUS HARVEST MOUSE. This large, brown harvest mouse is a Mexican species, which ranges northward into Arkansas and Missouri. It has crossed the Mississippi River and is extending its range northward across Mississippi; it may someday appear in western Kentucky.

Microtus chrotorrhinus (Miller). ROCK VOLE. This animal is quite similar to the meadow vole but differs in having a yellow nose. It often inhabits woody, rocky areas. Chiefly a northern species, it ranges southward along the crest of the Appalachians into North Carolina and Tennessee. It may occur at higher elevations in our southeastern mountains, but there are no records.

Rattus rattus (Linnaeus). BLACK RAT. This long-tailed, black rat has been erroneously recorded in Kentucky; the purported specimens are melanistic Norway rats. Actually, there may be local colonies of the black rat in Kentucky, but they have yet to be discovered.

Myocastor coypus (Molina). NUTRIA. This aquatic rodent, a native of South America, has been introduced into Kentucky. It weighs up to roughly 8 kg, and looks somewhat like a small, rat-tailed beaver, or a much-overgrown, round-tailed muskrat. After it was introduced, the species lingered on for a few years, but gradually declined. Probably there are no nutria left in the state today.

Canis rufus (Audubon and Bachman). RED WOLF. This small wolf (weight up to 36 kg) varies in color from gray-black to nearly black, except for the muzzle, ears, and the outer parts of the legs, which are tawny. This southern wolf, nearly extinct, is now known to occur only sparingly in the south-central United States. Two wild canids recently killed in western Kentucky were said to be this species. Unfortunately the specimens were not saved.

Bassariscus astutus (Lichtenstein). RINGTAIL. This animal is similar to a raccoon but is slimmer, has a longer tail, and lacks the black facial mask. It is native to the Pacific Coast, the southern Rockies, and arid Southwest (to east-central Kansas), and there are scattered records in the East. One in the collections of the University of Kentucky was taken by raccoon hunters in Bourbon County a number of years ago. We understand that the hunters saw another later in Franklin County, but it was not captured. It is probable that these animals had been released or had escaped from captivity.

*Mustela nivalis Linnaeus. LEAST WEASEL. This diminutive weasel is best recognized by its small size and its extremely short tail— only a little more than 25 mm in length. This is a northern animal, ranging southward across northern Illinois and Indiana, essentially all of Ohio, and in the higher Appalachians to North Carolina. It has been found at Cincinnati, Ohio, and probably occurs in Kentucky.

Felis concolor Linnaeus. MOUNTAIN LION; PANTHER. This is the only large, long-tailed cat in the eastern United States. At one time it was common in Kentucky, but there have been no valid records for some 75 years. Although one occasionally reads newspaper accounts of this animal in Kentucky, unfortunately these are probably based on escaped animals or a vivid imagination.

Bibliography

Ambrose, H. W., III
   1962   A comparison of *Microtus pennsylvanicus* home ranges as
          determined by isotope and live trap methods. M.S. thesis,
          University of Kentucky, Lexington. 91 pp.
Anderson, R. M.
   1948   Methods of collecting and preserving vertebrate animals.
          Natl. Mus. Canada Bull. 69. 162 pp.
Audubon, J. J., and J. Bachman
   1845– *Viviparous Quadrupeds of North America.* 3 vols., 1,107
   1854   pp.
Bailey, V.
   1933   Cave life in Kentucky. Am. Midl. Nat. 5(14):385–635.
Barbour, R. W.
   1941   Three new mammal records from Kentucky. J. Mamm.
          22(2):195–196.
   1942   Nests and habitat of the golden mouse in eastern Ken-
          tucky. J. Mamm. 23(1):90–91.
   1950   Notes on banded bats. J. Mamm. 31(1):350.
   1951   The mammals of Big Black Mountain, Harlan County,
          Kentucky. J. Mamm. 32(1):100–110.
   1952   Prairie vole, *Microtus ochrogaster*, in eastern Kentucky.
          J. Mamm. 33(3):398–400.
   1952   Animal habitats on Big Black Mountain in Kentucky.
          Trans. Kentucky Acad. Sci. 13(4):215–220.
   1952   'Possum. Kentucky Nat. 8(2):37–41.
   1956   Two new mammal records from Kentucky. J. Mamm.
          37(1):110–111.
   1956   *Synaptomys cooperi* in Kentucky, with description of a
          new subspecies. J. Mamm. 37(3):413–416.
   1956   A record of *Microsorex hoyi* from Kentucky. J. Mamm.
          37(3):438.
   1957   Some additional mammal records from Kentucky. J.
          Mamm. 38(1):140–141.
   1963   *Microtus*: A simple method of recording time spent in
          the nest. Science 141:41.
   1963   Some additional bat records from Kentucky. J. Mamm.
          44(1):122–123.

Barbour, R. W., and B. L. Barbour
1950   Some mammals from Hart County, Kentucky. J. Mamm. 31(3):359–360.
Barbour, R. W., W. H. Davis, and M. D. Hassell
1966   The need of vision in homing by *Myotis sodalis*. J. Mamm. 47(2):356–357.
Barbour, R. W., and W. H. Davis
1969   *Bats of America*. University Press of Kentucky, Lexington. 286 pp.
Barbour, R. W., and C. H. Ernst
1968   Forearm length and wingspread in *Myotis sodalis*. Trans. Kentucky Acad. Sci. 29(1–4):8–9.
Barbour, R. W., and L. Gale
1955   A study of the food of foxes in central Kentucky. Trans. Kentucky Acad. Sci. 16(4):102–104.
Barbour, R. W., and W. L. Gault
1953   *Peromyscus maniculatus bairdi* in Kentucky. J. Mamm. 34(1):130.
Barbour, R. W., and S. Hardjasasmita
1966   A preliminary list of the mammals of Rcbinson Forest, Breathitt County, Kentucky. Trans. Kentucky Acad. Sci. 27(3–4):85–89.
Barbour, R. W., C. T. Peterson, D. Rust, H. E. Shadowen, and A. L. Whitt, Jr.
1973   *Kentucky Birds: A Finding Guide*. University Press of Kentucky, Lexington. 306 pp.
Barbour, R. W., and C. E. Smith
1956   *Pitymys pinetorum carbonarius* in Harlan County, Kentucky. J. Mamm. 37(1):121.
Barr, T. C., and R. M. Norton
1965   Predation on cave bats by the pilot black snake. J. Mamm. 46(4):672.
Batson, J. D.
1965   Studies on the prairie vole, *Microtus ochrogaster*, in central Kentucky. Trans. Kentucky Acad. Sci. 25(3–4): 129–137.
Bole, B. P., and P. N. Moulthrop
1942   The Ohio recent mammal collection in the Cleveland Museum of Natural History. Sci. Publ. Cleveland Mus. Nat. Hist. 6(5):83–181.
Brauer, A., and A. Dusing
        Sexual cycles of the gray squirrel. Trans. Kentucky Acad. Sci. 22(1–2):16–27.

Burt, W. H.
1952 A *Field Guide to the Mammals*. Houghton Mifflin Co., Boston. 200 pp.

Call, R. E.
1897 Some notes on the fauna of Mammoth Cave, Kentucky. Am. Nat. 31:377–392.

Campbell, D. Y.
1951 A study of *Peromyscus* in the Inner Bluegrass region of Kentucky. M.S. thesis, University of Kentucky, Lexington. 53 pp.

Davis, W. H.
1959 Taxonomy of the eastern pipistrel. J. Mamm. 40(4): 521–531.
1963 Aging bats in winter. Trans. Kentucky Acad. Sci. 24(1–2):28–30.
1964 Fall swarming of bats at Dixon Cave, Kentucky. Bull. Nat. Speleol. Soc. 26:82–83.
1964 Winter awakening patterns in the bats *Myotis lucifugus* and *Pipistrellus subflavus*. J. Mamm. 45(4):645–647.
1967 A *Myotis lucifugus* with two young. Bat Res. News 8:3.

Davis, W. H., and R. W. Barbour
·1965 The use of vision in flight by the bat *Myotis sodalis*. Am. Midl. Nat. 74(2):497–499.
1970 Homing in blinded bats (*Myotis sodalis*). J. Mamm. 51(1):182–184.

Davis, W. H., R. W. Barbour, and M. D. Hassell
1968 Colonial behavior of *Eptesicus fuscus*. J. Mamm. 49(1): 44–50.

Davis, W. H., M. D. Hassell, and M. J. Harvey
1965 Maternity colonies of the bat *Myotis lucifugus* in Kentucky. Am. Midl. Nat. 73:161–165.

Davis, W. H., M. D. Hassell, and C. L. Rippy
1965 *Myotis leibi leibi* in Kentucky. J. Mamm. 46(4):683–684.

DeBlase, A. F., S. R. Humphrey, and K. S. Drury
1965 Cave flooding and mortality in bats in Wind Cave, Kentucky. J. Mamm. 46(1):96.

Dusing, A. A.
1957 The seasonal breeding cycles of the gray squirrel, *Sciurus carolinensis*, in Kentucky. M.S. thesis, University of Kentucky, Lexington. 45 pp.

Fassler, D. J.
1972 An additional record of the hoary bat in Kentucky. Trans. Kentucky Acad. Sci. 33(1–2):36.

1973 An additional evening bat from south central Kentucky. Trans. Kentucky Acad. Sci. 34(1–2):46.

Funkhouser, W. D.
1925 Wildlife in Kentucky. Kentucky Geological Survey, Frankfort. 385 pp.

Gale, L. R., and R. A. Pierce
1954 Occurrence of the coyote in Kentucky. J. Mamm. 35(2): 256–258.

Garman, H.
1894 A preliminary list of the vertebrate animals of Kentucky. Bull. Essex Inst. 26:1–63.

Gault, W. L.
1952 A survey of recent mammals of the Inner Bluegrass region of Kentucky. M.S. thesis, University of Kentucky, Lexington. 102 pp.

Genoways, H. H., and J. R. Choate
1972 A multivariate analysis of systematic relationships among populations of the short-tailed shrew (genus *Blarina*) in Nebraska. Systematic Zool. 21(1):106–116.

Goldman, E. A.
1910 Revision of the wood rats of the genus *Neotoma*. North Am. Fauna no. 31. Washington. 124 pp.
1918 The rice rats of North America (genus *Oryzomys*). North Am. Fauna no. 43. Washington. 100 pp.

Goodpaster, W., and D. F. Hoffmeister
1950 Bats as prey for mink in Kentucky cave. J. Mamm. 31(4):457.
1952 Notes on the mammals of western Tennessee. J. Mamm. 33(3):362–371.
1954 Life history of the golden mouse, *Peromyscus nuttalli*, in Kentucky. J. Mamm. 35(1)16–27.

Goslin, R.
1964 The gray bat, *Myotis grisescens* Howell, from Bat Cave, Carter County, Kentucky. Ohio J. Sci. 64:63.

Guilday, J. E., H. W. Hamilton, and A. D. McCrady
1971 The Welsh Cave peccaries (*Platygonus*) and associated fauna, Kentucky Pleistocene. Ann. Carnegie Mus. 43: 249–320.

Hall, E. R.
1951 *The American Weasels*. Univ. Kansas Publ., Mus. Nat. Hist., 4:1–466.
1962 Collecting and preparing study specimens of vertebrates. Univ. Kansas Mus. Nat. Hist. Misc. Publ. no. 30. 46 pp.

Hall, E. R., and E. L. Cockrum
1953 A synopsis of the North American microtine rodents. Univ. Kansas Publ., Mus. Nat. Hist., 5(27):373–498.
Hall, E. R., and K. R. Kelson
1959 *The Mammals of North America.* Ronald Press, New York. 2 vols., 1,083 pp.
Hall, J. S.
1961 *Myotis austroriparius* in central Kentucky. J. Mamm. 42(3):399–400.
1962 A life history and taxonomic study of the Indiana bat, *Myotis sodalis.* Reading Public Mus. Art Gal. Sci. Publ. 12:1–68.
1963 Notes on *Plecotus rafinesquii* in central Kentucky. J. Mamm. 44(1):119–120.
Hall, J. S., and N. Wilson
1966 Seasonal populations and movements of the gray bat in the Kentucky area. Am. Midl. Nat. 75(2):317–324.
Hamilton, W. J., Jr.
1930 Notes on the mammals of Breathitt County, Kentucky. J. Mamm. 11(3):306–311.
1943 *The mammals of Eastern United States.* Comstock Publishing Co., Ithaca, N.Y. 432 pp.
Handley, C. O., Jr.
1952 A new pine mouse (*Pitymys pinetorum carbonarius*) from the southern Appalachian Mountains. J. Washington Acad. Sci. 42(5):152–153.
1955 New bats of the genus *Corynorhinus.* J. Washington Acad. Sci. 45(5):147–149.
1959 A revision of the American bats of the genera *Euderma* and *Plecotus.* Proc. U.S. Natl. Mus. 110:95–246.
Hardin, J. W.
1967 Waking periods and movement of *Myotis sodalis* during the hibernation season. M.S. thesis, University of Kentucky, Lexington. 31 pp.
Harley, J. P.
1972 A survey of the helminths of the muskrat, *Ondatra z. zibethica* Miller, 1912, in Madison County, Kentucky. Trans. Kentucky Acad. Sci. 33(1–2):13–15.
Harvey, M. J.
1967 Home range, movements, and diel activity of the eastern mole, *Scalopus aquaticus.* Ph.D. dissertation, University of Kentucky, Lexington. 56 pp.

Harvey, M. J., and R. W. Barbour
1965   Home range of *Microtus ochrogaster* as determined by a modified minimum area method. J. Mamm 46(3):398–402.
Hassell, M. D.
1963   A study of homing in the Indiana bat, *Myotis sodalis*. Trans. Kentucky Acad. Sci. 24(1–2):1–4.
1967   Intra-cave activity of four species of bats hibernating in Kentucky. Ph.D. dissertation, University of Kentucky, Lexington. 66 pp.
Hassell, M. D., and M. J. Harvey
1965   Differential homing in *Myotis sodalis*. Am. Midl. Nat. 74:501–503.
Hooper, E. T., and E. R. Cady
1941   Notes on certain mammals of the mountains of southwestern Virginia. J. Mamm. 22(3):323–325.
Howell, A. H.
1910   Notes on the mammals of the middle Mississippi Valley, with description of a new woodrat. Proc. Biol. Soc. Washington 22:23–34.
1914   Revision of the American harvest mice (genus *Reithrodontomys*). North Am. Fauna no. 36. Washington. 97 pp.
1915   Revision of the American marmots. North Am. Fauna no. 37. Washington. 80 pp.
1918   Revision of the American flying squirrels. North Am. Fauna no. 44. Washington. 64 pp.
1929   Revision of the American chipmunks (genera *Tamias* and *Eutamias*). North Am. Fauna no. 52. Washington. 157 pp.
Humphrey, S. R., and J. B. Cope
1964   Movements of *Myotis lucifugus lucifugus* from a colony in Boone County, Indiana. Proc. Indiana Acad. Sci. 72:268–271.
1968   Records of the evening bat, *Nycticeius humeralis*. J. Mamm. 49(2):329.
Jackson, H. H. T.
1915   A review of the American moles. North Am. Fauna no. 38. Washington. 100 pp.
1928   A taxonomic review of the American long-tailed shrews (genera *Sorex* and *Microsorex*). North Am. Fauna no. 51. Washington. 238 pp.

Jackson, H. H. T.
1961 *The Mammals of Wisconsin.* University of Wisconsin
     Press, Madison. 504 pp.
Jegla, T. C.
1963 A recent deposit of *Myotis lucifugus* in Mammoth Cave.
     J. Mamm. 44(1):121–122.
Jegla, T. C., and J. S. Hall
1962 A Pleistocene deposit of the free-tailed bat in Mammoth
     Cave, Kentucky. J. Mamm. 43(4):477–481.
Jones, J. K., Jr., D. C. Carter, and H. H. Genoways
1973 Checklist of North American mammals north of Mexico.
     Occ. Papers Mus. Texas Tech. Univ. no. 12. 14 pp.
Kellogg, R.
1939 A new red-backed mouse from Kentucky. Proc. Biol.
     Soc. Washington 52:37–40.
Krutzsch, P. H.
1954 North American jumping mice (genus *Zapus*). Univ.
     Kansas Publ., Mus. Nat. Hist., 7(4):349–472.
Lyon, M. W., Jr.
1936 Mammals of Indiana. Am. Midl. Nat. 17(1):1–384.
Metzger, B.
1956 Partial albinism in *Myotis sodalis.* J. Mamm. 37(4):546.
Miller, G. S., Jr.
1897 Revision of the North American bats of the family
     Vespertilionidae. North Am. Fauna no. 13. Washington.
     80 pp.
Miller, G. S., Jr., and G. M. Allen
1928 The American bats of the genera *Myotis* and *Pizonyx.*
     U.S. Natl. Mus. Bull. no. 144. Washington. 215 pp.
Miller, G. S., Jr., and R. Kellogg
1955 List of North American recent mammals. U.S. Natl.
     Mus. Bull. no. 205. Washington. 954 pp.
Neel, J. K.
1938 Lower Howard's Creek: a biological survey. M.S. thesis,
     University of Kentucky, Lexington. 233 pp.
Nelson, E. W.
1909 The rabbits of North America. North Am. Fauna no. 29.
     Washington. 314 pp.
Osgood, W. H.
1909 Revision of the mice of the genus *Peromyscus.* North
     Am. Fauna no. 28. Washington. 285 pp.
Packard, R. L.
1969 Taxonomic review of the golden mouse *Ochrotomys*

nuttalli. Univ. Kansas Mus. Nat. Hist., Misc. Publ. 51: 373–406.

Patterson, P., and W. H. Davis
  1968 Variations in size among adult *Eptesicus fuscus.* Trans. Kentucky Acad. Sci. 29(1–4):1–4.

Patterson, A. P., and J. W. Hardin
  1969 Flight speeds of five species of vespertilionid bats. J. Mamm. 50(1):152–153.

Pearson, E. W.
  1962 Bats hibernating in silica mines in southern Illinois. J. Mamm. 43:27–33.

Phillips, R. E., and R. E. Mumford
  1957 A record of *Myotis austroriparius* for Kentucky. J. Mamm. 38(4):515.

Putnam, F. W.
  1874 Remarks on the Mammoth Cave and some of the animals. Bull. Essex Inst. 12(6):191–200.

Quay, W. B., and J. S. Miller
  1955 Occurrence of the red bat, *Lasiurus borealis,* in caves. J. Mamm. 36(3):454–455.

Rafinesque, C. S.
  1818 Am. Monthly Mag. 3:445–446.
  1832 Atlantic J. 1:61.

Rippy, C. L.
  1965 The baculum in *Myotis sodalis* and *Myotis austroriparius.* Trans. Kentucky Acad. Sci. 26:(1–2):19–21.
  1967 The taxonomy and distribution of the short-tailed shrew, *Blarina brevicauda,* in Kentucky. M.S. thesis, University of Kentucky, Lexington. 84 pp.

Rippy, C. L., and M. J. Harvey
  1963 Comparative behavioral characteristics of six genera of mice. Trans. Kentucky Acad. Sci. 24(1–2):5–8.
  1965 Notes on *Plecotus townsendii virginianus* in Kentucky. J. Mamm. 46(3):499.

Robinson, T., and F. W. Quick
  1965 The cotton rat in Kentucky. J. Mamm. 46(1):100.

Robinson, T. S., K. T. Reichard, and W. L. Thomas
  1965 New distributional records of the prairie vole in Kentucky. Trans. Kansas Acad. Sci. 68(1):204–205.

Shackelford, N.
  1966 Eastern chipmunk feeding on a starling. J. Mamm. 47(3):528.

Shadowen, H. E.
  1951 An ecological survey of the genus *Microtus* in the Inner
        Bluegrass region of Kentucky. M.S. thesis, University of
        Kentucky, Lexington. 40 pp.
Smith, C. E. T.
  1956 The genus *Pitymys* in Kentucky. M.S. thesis, University
        of Kentucky, Lexington. 51 pp.
Tichenor, T. C.
  1954 A study of the mammals of Jefferson County, Kentucky.
        M.S. thesis, University of Kentucky, Lexington. 67 pp.
Wade, J. K.
  1970 Home ranges, activity patterns, and nesting habits of
        Blarina brevicauda. M.S. thesis, University of Kentucky,
        Lexington. 67 pp.
Wallace, J. T.
  1969 Some notes on the growth, development and distribution
        of *Ochrotomys nuttalli* (Harlan) in Kentucky. Trans.
        Kentucky Acad. Sci. 30(1–2):45–52.
  1971 New records of *Zapus hudsonius* (Zimmerman) from
        Kentucky. Trans. Kentucky Acad. Sci. 32(3–4):65–69.
Wallace, J. T., and R. Houp
  1968 Marginal record of *Parascalops breweri* (Bachman) from
        Kentucky. Trans. Kentucky Acad. Sci. 29(1–4):9.
Welter, W. A., and D. E. Sollberger
  1939 Notes on the mammals of Rowan and adjacent counties
        in eastern Kentucky. J. Mamm. 20(1):77–81.

Index

grizzly bear, 305
groundhog. *See* woodchuck
ground squirrel, 307. *See also* eastern chipmunk
guano bat. *See* Brazilian free-tailed bat

hairy-tailed mole, PLATE 4, 45-48
hispid cotton rat, PLATE 20, 189-92
hoary bat, PLATE 10, 103-5
house bat. *See* big brown bat
house mouse, PLATE 25, 224-28

Indiana bat. *See* Indiana myotis
Indiana myotis, PLATE 7, 76-81
insectivores, 26-53

jumping mice, 228-37

Keen's bat. *See* Keen's myotis
Keen's myotis, PLATE 6, 72-75

Lasionycteris noctivagans, PLATE 8, 85-89
Lasiurus borealis, PLATE 9, 99-103; *L. b. borealis*, 99
Lasiurus cinereus, PLATE 10, 103-5; *L. c. cinereus*, 103
Lasiurus intermedius, 307
Lasiurus seminolus, 307
least shrew, PLATE 3, 41-44
least weasel, 260, 309
Leib's bat. *See* small-footed myotis
lemming mouse. *See* southern bog lemming
Lepus americanus, 307
little brown bat. *See* little brown myotis
little brown myotis, PLATE 5, 60-65
long-tailed shrew, 306
long-tailed weasel, PLATE 28, 260-62
Lontra canadensis, PLATE 30, 274-78; *L. c. canadensis*, 274-75; *L. c. interior*, 275
Lutra, 275
Lynx rufus, PLATE 31, 278-81; *L. r. rufus*, 278

mammoth, 305
Mammuthus, 305
Marmota monax, PLATE 14, 139-43; *M. m. monax*, 139
marsh rice rat, PLATE 17, 165-68

marsupials, 19-26
masked shrew, PLATE 1, 28-32
meadow jumping mouse, PLATE 25, 229-33
meadow mouse. *See* meadow vole
meadow vole, PLATE 22, 199-204, 308
Mephitis mephitis, PLATE 30, 269-74; *M. m. nigra*, 269
mice, 168-89, 196-212, 224-37, 308
Microsorex thompsoni, 306
Microtus chrotorrhinus, 308
Microtus ochrogaster, PLATE 22, 189, 204-8; *M. o. ochrogaster*, 204-5; *M. o. ohionensis*, 205
Microtus pennsylvanicus, PLATE 22, 199-204; *M. p. pennsylvanicus*, 201
Microtus pinetorum, PLATE 23, 201, 208-12; *M. p. auricularis*, 209; *M. p. carbonarius*, 209
Microtus xanthognathus, 305
mine rat. *See* Norway rat
mink, PLATE 29, 262-65
moles, 44-53, 306
mountain lion, 309
muskrat, PLATE 23, 212-17
Mus musculus, PLATE 25, 224-28
Mustela frenata, PLATE 28, 260-62; *M. f. noveboracensis*, 260
Mustela nivalis, 260, 309
Mustela vison, PLATE 29, 262-65; *M. v. mink*, 263; *M. v. vison*, 263
Myocastor coypus, 308-9
Myotis austroriparius, PLATE 5, 65-68
Myotis grisescens, PLATE 6, 69-72
Myotis keenii, PLATE 6, 72-75; *M. k. septentrionalis*, 72
Myotis leibii, PLATE 7, 81-85; *M. l. leibii*, 81
Myotis lucifugus, PLATE 5, 60-65, 75; *M. l. lucifugus*, 60
Myotis sodalis, PLATE 7, 76-81

Napaeozapus insignis, PLATE 25, 233-37; *N. i. roanensis*, 233
Neotoma floridana, PLATE 21, 192-96; *N. f. illinoensis*, 193; *N. f. magister*, 193
New England cottontail, PLATE 12, 127-29
New World rats and mice, 162-221